WILSON
Livestock Express
OPERATED BY
LIVESTOCK EXPRESS
SOUTH ST. PAUL
MINNESOTA
IA.-43492
WIS.-8512
MO.-T-19423
Y65-125

BIG RIGS of the 1960s

RON ADAMS

MOTORBOOKS

First published in 2004 by Motorbooks, an imprint of MBI Publishing Company, 380 Jackson Street, Suite 200, St. Paul, MN 55101-3885 USA

ISBN 0-7603-1618-X

Edited by Peter Bodensteiner
Designed by Mandy Iverson

Printed in China

On the front cover:
Clockwise from top right: Chevron Oil Co. delivered their gasoline in a Fruehauf tank trailer that was powered by a V-series International rig. A stack on this V-series tractor indicates it is probably diesel powered; a good way to show off a brand new trailer is to match it with a brand new tractor. Such is the case with the brand new Hobbs Vanguard reefer trailer that was introduced in 1964. A nice colorful and chromed K100 Kenworth makes a good partner for the new truck. Many Texas reefer haulers used this model trailer; this early 60s International Emeryville is decked out with a nice color scheme and a matching Great Dane reefer trailer. This is one of the many sharp looking rigs that ran in the reefer division for Transamerican Freight Lines Inc. of Detroit, Michigan; the Monfort Packing Co. of Greeley, Colorado introduced their over-the-road fleet in the fall of 1970. This 1970 W900 Kenworth was powered by a 425-horsepower Caterpillar diesel engine and pulled a Timpte reefer trailer; the White Freightliner is typical of the type of rigs that ran in the western states. Dromedary bodies were common on a lot of cab-over-engine tractors with long wheelbases. Often the money made from the freight on the dromedary body paid for the expenses of the trip, while the money made from the freight in the trailer was profit.

On the frontis: Livestock Express of South St. Paul, Minnesota, depended on this 931 C to get loads of livestock safely from farm to market in a Wilson livestock trailer. *Wilson Trailer Co.*

On the back cover: Along with Kenworth's new K-100 series came the new W-900 series. This very handsome-looking truck had a wide-front radiator tilt hood. The wide-front design was to accommodate bigger engines. The standard engine was a Cummins NH-220. *Kenworth Truck Co.*

Marmon trucks were made in Garland, Texas. This one was owned by Doug Parker of Dearborn Heights, Michigan, and leased to Michigan Transportation Co. also of Dearborn, Michigan. The 1966 model is loaded down with steel riding on eleven axles. The "China Clipper" is ready to roll. *Detroit Diesel*

About the author: Ron Adams is a lifelong big truck enthusiast who has spent many years amassing an archive of vintage truck photos. This is his fourth book for MBI Publishing, following *100 Years of Semi Trucks, Big Rigs of the 1950s,* and *Mack Trucks.* He lives in Lenhartsville, Pennsylvania.

CONTENTS

PREFACE

Since the earliest days of trucks, every truck that was produced had a model letter or model number. In the early years the manufacturers did not have big lines to offer, but by the 1920s some models started to have different letters or numbers. In the later 1920s, Mack started a "B" series, which consisted of the BG, BJ, BB, BC, BF, BL, BM, BX, and BQ. Although a lot of them looked similar, they were in different weight classes, which meant that they either had a light or a heavy rear end or the front axle was lighter or heavier. The International "D" series of the 1930s followed the same pattern, as did the "K" series of the 1940s. The engine size was also reflected by a model letter or number. Through the 1950s and 1960s, Mack offered 68 different "B" series model numbers from the B-13, which was the smallest light-duty model, up through the B-873. Each of the numbers in the "B" series had something different from the others. All the manufacturers followed this pattern, and you will notice as you look through this book that a lot of them introduced models in the 1950s that they carried over, with improvements, into the 1960s. Of course, each truck design builds on what came before, so the sharp enthusiast will spot design features and accessories on these pages tracing back to the truck's introduction.

DEDICATION

Wanting to drive a truck, either professionally or just to feel the experience, is a dream that many young people have at one time or another. What inspires this dream is anybody's guess. Maybe it's the sight of those big-looking beasts with both stacks blowing black smoke. Maybe it's because their father drove a truck. Maybe it's because these youths were around trucks all their lives. Or maybe it's a vision of themselves as big tough guys behind the wheel of one of those big tough trucks. Whatever the reason may be, for some it's just a dream. For others, it's a dream come true. For the four Follweiler brothers, I guess it's something that just happened through opportunity.

This book owes its creation in part to the four sons of George and Barbara Follweiler: George Jr., Wayne, Warren, and Archie. There were also three daughters: Grace, Dorothy, and Lila. George Follweiler started a little fruit growing business in Stony Run, Pennsylvania, around 1920. As the years went on, his little fruit business grew bigger and bigger. At the same time, his family grew bigger and bigger. By 1931, he had seven children with Archie being the youngest.

Through the 1930s and 1940s, the once little fruit business grew to seven farms totaling over 1,000 acres of fruit trees, berries, fruits, and vegetables. The four brothers never had many toys to play with because they were out in the fields and orchards with the many other workers using shovels, hoes, rakes, and tree pruners. As they became teenagers, they learned to operate the heavier farm equipment. Not long after that, they started to drive the trucks that their father used to haul fruit to markets in Scranton, Philadelphia, New York City, and several other cities.

The 1930s trucks they drove were Chevrolet, Dodge, International, and REO straight trucks. The 1940s trucks were K-5 International, D-35 International, K-7 International, and K-7 International Sleeper. The seeds of their truck driving careers were planted. By the early 1950s, the brothers were all married with their own families. One by one the brothers started to leave, going out on their own as truck drivers.

The fruit farm continued, and the brothers helped out during the busy seasons until the business was sold in 1967. I started working on the fruit farm in 1958 at the age of 12 and got to know the brothers and their trucks. They would often park at their father's packing house garage to do some work on their trucks or trailers and tell me stories. At this time I was about seventeen years old and still in high school.

George Jr. was the oldest of the boys. He first drove for his brother Warren, driving one of his 1951 LF Macks with a Lanova diesel. The truck was leased to Interstate Truck Service of Wheeling, West Virginia. George drove for Warren for a while and then quit. He then drove for Peters Bros. in Lenhartsville, Pennsylvania, hauling livestock from the Midwest into Allentown, Pennsylvania. After working for Peters Bros. for several years, he started driving for Edgar Black of Skippak, Pennsylvania, hauling livestock again. He drove a B-61 Mack with a Fruehauf straight-floor cattle trailer. A truck accident cut his life short in July 1970, when he lost control on I-78 west of Route 22 after a heavy rainstorm.

Wayne, the next brother in line, began his driving career around 1955. He started with Revel Oliver, who owned a small fleet of trucks. Wayne drove a B-61 Mack, pulling an Ohio body open top trailer. The truck was also with Interstate Truck Service, running from Philadelphia, Pennsylvania, and the New York City, Newark, and Jersey City area to several cities in Ohio hauling all kinds of mixed freight. He also drove an H-61 "Cherry Picker" Mack. After seven years with Revel Oliver, he quit and started driving for Burgmeyer Bros. in Reading, Pennsylvania, in 1962. Most of the trucks he drove for them were Whites. He gave them nineteen years of driving until the company went out of business in 1981. Next he drove for Wetereau Transportation Co., also in Reading, for thirteen years. After that, he went to Super Value for seven years until he retired in 2001. Wayne gave 46 years to his driving career and did it well.

After Wayne, Warren was next in line. His first driving job was at the wheel of a 1951 Dodge with a Flathead 6, for a man in Allentown, Pennsylvania. He also had his truck on with Interstate Truck Service. After a short time on the Dodge he quit and bought his own rig, a 1954 R-200 International that was painted orange and green, and pulled his own 32-foot Trailmobile tandem trailer. He purchased the truck from Berman Leasing Inc., where it had previously been leased by the Atlas Powder Company. After six months the engine went out in Somerset, Pennsylvania. He switched to a red GMC for Howard Newell, the terminal manager for Interstate Truck Service at the Akron, Ohio, office. Warren next bought a 1951 and a 1952 LF Mack, both with Lanova diesels. He added a 1957 "Cherry Picker" Mack with a 165 horsepower engine and a brand new 1961 B61 Mack, bringing his fleet of trucks to four—two were on with Interstate Truck Service and two were on with Travelers Motor Freight. (He also had a "C" model Autocar.) A brand new 1962 F700 Mack replaced the whole fleet, but he only had it six months before it was stolen in New York City. Two "G" model Macks, a 1961 and 1962, filled the void. The 1961 was a single axle which he converted to a tandem axle, and the 1962 was a tandem axle. He pulled for Sam Winer Motors in Cleveland, Ohio. A used Fruehauf Reefer Trailer, once owned by Midwest Emery Freight System, came next, which he leased on with First Scott Street Meat Packing Company in Detroit, Michigan. He also made a trip to the west coast with his "G" model delivering a load of safes to San Diego, California.

After selling both "G" models in 1966, Warren moved his family to Akron, Ohio, and started driving for Yankee Lines Inc. Three months later they were taken over by Mason-Dixon Lines. After 16 years he left and bought a 1979 Freightliner COE with a 400 horsepower Cummins diesel, pulling his own Reefer Trailer for Rowley Interstate Inc. of Dubuque, Iowa. Next was a 1985 LTL 9000 Ford, which was eventually replaced by a new 1989 Volvo with a 3406 CAT diesel, replaced in turn by a new 1995 Volvo, also with a 3406 CAT diesel. His last truck was a 1998 Volvo with a 500 horsepower Detroit Diesel, after which he retired following 54 years of driving. Fifty-four years behind the wheel of a truck is a long time. Like all truckers, Warren had his bad times and good times, but the good outweighed the bad. Being the tough guy, he took it all in stride and made a tough job a lot of fun.

Archie, the youngest of the four brothers, got his first driving job with James Delong Grain Company in Orefield, Pennsylvania. He drove a 1950 DC 75 Autocar with a flat windshield and the oil pan guard underneath, hauling grain from the Orefield granary to Long Island—always overloaded. He said it was no problem for that Cummins diesel. Next he drove a K-8 International for George W. Brown Inc. in Allentown, Pennsylvania. He also drove an "R" model cab-over engine International, hauling a load of apples out to Washington state. After a while with Brown, he started to drive for Max Rabel, who had a fleet of trucks leased to Interstate Truck Service of Wheeling, West Virginia, and to A-C-E Transportation Co. Inc. of Akron, Ohio. One of the trucks he drove was a B61 Mack (see picture on page 117 of *100 Years of Semi-Trucks*) for A-C-E Transportation Co. He also drove an H61 Mack leased with Interstate Truck Service. In 1958 he drove a DC 75 Autocar with a 200 Cummins for a man named Clafflin, who also leased his truck with Interstate Truck Service.

In the early 1960s Archie bought his first truck, a used DC 75 Autocar with a 165 horsepower Cummins. It had a homemade muffler on it that gave a sound you wouldn't believe. He bought a second DC 75 Autocar with a 175 horsepower Cummins. Archie drove the second one and Bobby Reichard drove the first one, both leased to W. H. Kilbourne Inc. of Akron, Ohio. Selling the Autocars, he bought an ex-Roadway Express H67 Mack, which he had leased to C.A. Givens Inc. In 1968 he got out of the trucking business and bought two service stations in Fogelsville, Pennsylvania. Several years later he sold them and bought the Kempton Hotel and Restaurant in Kempton, Pennsylvania. He and his partner also ran Kempton Excavating Co. for a few years. Archie still owns the hotel and restaurant today. He wasn't in the trucking business as long as his brothers, but he was a true dedicated trucker, always as tough as the trucks he drove. He was and is a great guy.

Since I've known the four brothers for 45 years and their trucking careers started over fifty years ago, I want to dedicate this book to the "Truckin' Brothers Four." These four brothers worked hard, learned a lot, and made the trucking business work. Their boss at Interstate Truck Service liked them very much and credited them as hard working Dutchmen. They put on many miles running from the Northeast to the Midwest over the Pennsylvania Turnpike and Route 22. At that time it was their job, but today it is only a memory. So to the Truckin' Brothers Four—this one's for you.

This DC-10064 is the main attraction in this interesting photograph. The DC 100 model was Autocar's offering to the heavy-duty western truckers. C. B. Edmunds Transportation was a heavy hauler based in Casper, Wyoming, and Plenttwood, Montana. Oilfield hauling and rigging was its main business. All the rigging equipment is on the tractor as needed, along with the unknown brand name Tri-Axle Lowbed Trailer. The photo was taken at Rick's Truck Stop, assumed to be in Denver, Colorado. In the background, the tanker trailers and the tractor fleet belong to Ward Transport Co., also of Denver. *Birlauf & Steen Inc.*

CHAPTER 1

AUTOCAR

Being a good tough model, the DC and the "A" series of the 1950s carried into the 1960s. The DC used all steel components while the "A" series was of lightweight aluminum components. There were a few different types in the DC series. The DC 75 was the axle back model; the DC 100 was the axle forward model—which was designed mainly for the western trucker—and the DCV model was the short hood version. The "A" series, which started in 1958, had the same kind of designs. Its all-aluminum construction was designed to save weight for bigger payloads. Both series saw the different models across all industries.

Autocars were very popular in the construction industry because of their toughness and durability. The "S" series, a tilt-hood design, was supposed to debut in the later 1960s, but for unknown reasons never went into production. Another late 1960s introduction was the A64F, a tilt-hood design that accommodated a big V12-71N diesel engine. It was the 1969 ATA show truck, later owned by Andy Henderson of Monroe, New York. Since day one, Autocars have proven themselves enough times to bring them through their first seventy years.

Another Autocar, a DC 102, based in Upton, Wyoming, for Young Trucking Co., doing the same kind of work. This single axle model is on a 220-inch wheelbase and is powered by a 220-horsepower Cummins diesel engine with a 5x4 transmission. The front axle is rated for 11,000 pounds and the rear axle for 23,000 pounds riding on 18x22.5 flotation tires. The dual chrome stacks and chrome radiator shell brighten up the appearance of the rig. *Autocar Trucks*

Autocar trucks were used in all segments of trucking. This 1962 DC-75 is owned by Ray Mathes of Louisville, Kentucky, and leased to Coldway Carriers Inc. of Clarksville, Indiana, pulling a Fruehauf Reefer Trailer. This chrome loaded rig could be seen every Sunday afternoon or evening heading east on Interstate 78 to make east coast deliveries in the mid 1960s. The photo was taken in 1968 at the Roadway Truck Stop near Carlisle, Pennsylvania. *Ron Adams Collection*

The DCV 72 was introduced in the early 1960s. This plain Jane model was the shortest tractor in the Autocar line. It was designed to pull 40-foot trailers and stay within the legal 50-foot length in the 1960s. *Autocar Trucks*

Another 1962-era Autocar, a DC that was used by Via Contractors Inc. of Richmond, Virginia, to haul road-building equipment on a lowbed trailer. The heavy metal construction-type fenders on the front and full fenders over the rear provide protection for the cab and other parts of the tractor. *Autocar Trucks*

Another DCV 72 T, pulling double Michigan-style tank trailers for Farmers Petroleum Coop. of Lansing, Michigan. It specs out to a 132-inch wheelbase with power from a V-8 265 horsepower diesel engine riding on 11.00x20 tires. *Autocar Trucks*

The same model but with a tandem axle. The DCV 7264 TL is on a 157-inch wheelbase with power from a V-8 265-horsepower diesel engine with a 10 speed transmission riding on 11.00x22.5 tires. It is pulling a petroleum tank trailer for Ohio Oil Co. in Findley, Ohio. This model was also available with an integral sleeper cab. *Autocar Trucks*

The "A" series, the all-aluminum lightweight series, started in 1958. The "A" series was available in several models, such as this A102T. This seemed to be the most popular model for the western trucker pulling doubles. *Autocar Trucks*

This A102 is pulling a set of double milk tankers for Kings County Truck Lines of Tulare, California. Notice that the three middle lights on top of the cab have green lenses. *Brian Williams*

Another model in the "A" series, the A75, pulling a reefer trailer for an unknown carrier. Integral sleeper cabs were also available on the series. The bumper is decked out with the typical 1960s license plates for all the states in which it traveled. *Brian Williams*

Here we have another version of the A75, pulling a tank trailer for Provost Cartage Ltd. of Montreal, Quebec, Canada, in 1967. Notice the small waving hand decal on the left fender. *Provost Cartage*

Another A75 model of the 1960s era for the Timken Roller Bearing Co. of Ohio. The driver of this truck has the comfort of an air conditioner. *Ron Adams Collection*

This "A" series West Coaster hauled meat products or produce in the Great Dane Reefer Trailer for Walter Strauss and Son of Fort Worth, Texas. Notice the fender mount dual headlights. Even though integral sleeper cabs were still offered, this owner chose an add-on box sleeper. *Neil Sherff*

Autocar trucks seemed to be very popular in the logging business. This "A" series was truck number 20 in the fleet for Alan Kramer Logging Co. of Coos Bay, Oregon. It hauled logs with a Peerless logging trailer. *Autocar Trucks*

Another "A" series model, an AU7064T from the 1964 era. This model was 12,300 pounds with standard components or 11,292 pounds with weight-saving components. It was offered with a 220-horsepower diesel engine as standard or a weight-saving compact V-8 235 horsepower diesel engine. The Mobil Oil Co. chose this one for pulling gasoline tank trailers. *Mobil Oil Co.*

Autocar trucks were made in all sizes. This heavy duty hauling model was a DC 9966 with a GVW of 73,000 pounds. It's anybody's guess what this particular truck hauled, but you can bet that this giant got the job done. The "66" in the model number means it is a six wheeler with all six wheels driving. *Autocar Trucks*

I have no information about this model other than that it is known as the "S" series. The license plate shows Colorado 1969. My guess is that this was a prototype that never went into production but marked the beginning of the restyling of the A64 model of the same era. *Ron Adams Collection*

This was Autocar's 1969 ATA show truck, an A64F with the new tilt-hood design. It was powered by a 12V-71N Detroit Diesel engine. It was designed for both on/off highway jobs. The transmission is a 13-speed Fuller RTO-1213. *Autocar Trucks*

This model 257 of the early 1960s era was in the Allied Chemicals fleet based in Camden, New Jersey. The standard engine was an NH-220 diesel. The model 60J cab was used on both the 257 and the model 260. *Brockway Trucks*

CHAPTER 2

BROCKWAY

In 1956 Brockway was purchased by Mack Trucks Inc. During this time, Brockway started manufacturing oilfield trucks. The company also introduced a new model 258, which was a short hood and cab design. These models continued into the 1960s. In 1962 Mack introduced their new "F" series cab-over engine. Not long after that, this same style cab-over engine model, containing the same Mack engine, was wearing the Brockway name, and also the Huskie Dog emblem.

Brockways continued to sell well in the northeast. On conventional Brockway models, Mack retained Brockway's own cab designs until the later 1960s. At that time Brockway started to use the Mack "R" series cab on their conventionals. This led to some new models that went into the 1970s. By keeping the Brockway name, Mack made production easy for its engineers: they could create one cab design and get two new trucks out of it by using the Mack and Brockway badges. The two different dogs trotted the same territory side by side, one just as tough as the other.

The New York countryside makes a nice background as this early 1960s 258 TL Brockway and Strick trailer pose on a sunny day. The standard engine for this model was an NH-220 diesel. The 87-inch BBC (from bumper to back-of-cab) worked well for pulling 40-foot trailers. Penn Yan Express Inc. of Penn Yan, New York, was a big Brockway user. *Penn Yan Express*

Brockway trucks hauled their share of heavy loads. This model 260 had no problem handling the job for D. Ali Trucking for United States Concrete Pipe Co. A spread-axle flatbed trailer was used to balance out the weight. *Ron Adams Collection*

The town of Walworth needed a helping hand, so this 260 was chosen to do the hauling chores. It probably was set up as a snow plow for winter work. *Brockway Trucks*

The 300 series was introduced in the mid-1960s. Here we see a fleet of six model 358s all equipped with City dump bodies from Lampl Asphalt Paving Co. of Dravosburg, Pennsylvania. The new 300 series came with a restyled grille. *Ron Adams Collection*

The 358 also came in a single-axle version. Tractor number 8 was in the Ithaca Delivery Inc. fleet of Ithaca, New York. *Brockway Trucks*

In 1962, Mack introduced their new "F" series cab-over. This same cab was also used on the new Brockway series cab-over, since Mack owned Brockway. This model N457-TL is powered by an NH-250 Cummins diesel with a 12-speed Spicer transmission. The Brockway tractor and Trailmobile trailer were owned by Yankee Lines Inc. of Akron, Ohio, "Route of the Minute Men," as their slogan says. The picture was taken at the Trexler Oil Truck Stop in Kuhnsville, Pennsylvania, in 1966. *Bob Parrish*

After Mack purchased Brockway, they started to use the new "F" series cabs. Mack put the Brockway name on it and called it the 400 series. This model 459 and the Trailmobile trailer were leased to the Brixite Manufacturing Co. in South Kearny, New Jersey. Engine availability went up to 335 horsepower with either 5-, 10-, or 15-speed transmissions. *Bob Parrish*

Here we have another 459 that is hauling a 64,000-pound, 14 1/2-foot-high water tube boiler on a tri-axle lowbed trailer. Kreitz Motor Express Inc. of Reading, Pennsylvania, is ready for the 119-mile trip from Lebanon, Pennsylvania, to Somerville, New Jersey. The company averaged 20 miles an hour on the trip. *Heavy Duty Trucking 9/70*

The 400 series was also available in a single-axle version. The model 459 was owned by longtime horse transporter, Ralph G. Smith Inc. of West Chester, Pennsylvania. This lucky horse show tractor transported show horses all over the country in the Fruehauf trailer. Engine and transmission availability for the single axle is the same as for the tandem axle model. *Ron Adams Collection*

A model 457-TL showing off its custom paint job. *Brockway Trucks*

A model 550 L designed for local delivery work as a straight truck, even though they were available in a tractor version with tandem or a single axle. Power came from a Detroit Diesel, Caterpillar, or Cummins engine. *Brockway Trucks*

Standing proudly on display is this model 359 decked out with plenty of shining chrome. This 90-inch B.B.C. model could accommodate V-6 and V-8 engines up to 335 horsepower. Access to the engine is provided by the raised cab position and step side fenders. *Brockway Trucks*

This model 359 is owned by Purdie Express Lines Inc. of Brockport, New York. The trailer is a Gindy with dual hoppers for hauling bulk commodities. The dual stacks could mean that the tractor is powered by a 318-horsepower Detroit Diesel engine. Philip J. Boncaro, executive vice president of the company, explains the functions of the hopper trailer to General Manager Bob Viscardi. *Ron Adams Collection*

Belbey Transfer Co. of Kearny, New Jersey, was a heavy hauler and machinery mover in the northeast. A model 359 Brockway was the source of power to move this heavy generator on the lowbed trailer. As seen on the sign, this company was established in 1905. *Brockway Trucks*

A model 361 in the testing stage, trying to prove that it is capable of doing what was expected of it. The flatbed trailer is carrying the test load. The engine range was from 230 to 320 horsepower. *Brockway Trucks*

A Brockway Model 761 LL dump truck. *Brockway Trucks*

This model 761 TL with its custom paint design looks similar to the 361 except for the model 78 cab and the redesigned headlights. The model 761 came with Cummins, Caterpillar, or Detroit Diesel power. Available engines ranged from a Cummins NH-230 horsepower up to a Detroit Diesel 8V-92T with 430 horsepower. Fuller and Dana transmissions were available, as were Allison automatics. *Brockway Trucks*

The Spartan model debuted in 1960. This man stands proud on the running board of his 70 series Spartan hauling a load of logs on the Nabors logging trailer. Notice that this tractor still uses the small round mirror instead of the west coast type. *Fred McPhearson*

CHAPTER 3

CHEVROLET

Owned by GMC, Chevrolet held their own in production. Both Chevrolet and GMC trucks looked almost identical with a few minor design changes. In 1960, a new line of cab-over engine models came into production. The "A" model series 80 was introduced in 1965, and was available with both gas and diesel engines. Another new design was the non "A" model series 80, also available in both gas and diesel. All these models were available in both 4x2 and 6x4 setups. In 1969 Chevrolet introduced their first-ever high tilt cab-over-engine tractor, the Titan 90. It had the biggest windshield of any Chevrolet model, and engines up to the big V-12. The 80 series conventional was available in both the short hood and the long hood to accommodate the bigger engines. At this time GMC and Chevrolet trucks shared the same styling.

Another model for Chevrolet was the Viking. This one is equipped with a 10x7, 4 yard capacity, and Black Diamond dump body. *Black Diamond Trailer Co.*

This new tilt cab model was also introduced in 1960. United Van Lines of St. Louis, Missouri, owned this tilt cab tractor and Dorsey moving van trailer. The models in the early half of the 1960s sported the double headlights. This model was very popular with a lot of the moving companies. *United Van Lines*

This tandem-axle tilt cab was based in Ohio. Here it pulls an Ohio spread axle trailer in 1962. Notice the west coast mirrors. *Neil Sherff*

This 1961 series 60 was in the middle-weight class. It was a city delivery truck for Centralia Cartage Co. of Centralia, Illinois. The van body was built by Hewitt-Lucas. *Hewitt-Lucas Body Co.*

There was a change in the grille design in 1962 when the series 60 went from double to single headlights. This new model was known as the Jobmaster. This one was sold by Hicks Chevrolet Company to the Alabama Oil Co. in Decatur, Alabama. They were the distributor for Cities Service Gasoline, which no longer exists. *Hicks Chevrolet*

Here we have a series 80, which was the heavy-duty model. The new grille featured single headlights and three small upright bars. The truck was owned by the Vizpah Container Co. in Virginia. *Harry Patterson*

This is an "L" series of 1964, with a Brown delivery body. The standard engine was a 292 six cylinder, but there was an optional 327 V-8. The wheelbase went from 133 inches to 197 inches. Automatic transmissions were also optional. *Ron Adams Collection*

In 1964, there was another grille change. The three short upright bars were eliminated and "Chevrolet" became individual letters. Homer Taylor was the owner of this "C" series tractor, pulling a Trailmobile moving-van trailer for Mayflower Transit Co. of Indianapolis, Indiana. A sleeper box was added for resting comfort. *Harry Patterson*

Another LCF 80 series as a tractor with an add-on sleeper box. The standard engine was the 348 V-8 with the 409 V-8 as an option. These models seemed to be favorites with moving companies. This one was leased to Engel Bros. World Wide Moving Co. in Bellvedere, New Jersey. *Harry Patterson*

For 1965, yet another new model appeared—the "A" model, series 80. This diesel model was powered by the 6V-53N Detroit Diesel. Tom Hicks Transfer Co. of Harvey, Louisiana, used this one for heavy hauling. *Tom Hicks Transfer Co.*

These two model "A" series 80 tractors for Anchor Motor Freight are practically new. The picture was taken in January 1965 at the Deepwater Truck Stop in Deepwater, New Jersey. These rigs are delivering new 1965 Buicks out of Wilmington, Delaware. *Bob Parrish*

The low-tilt cab tractors of the Spartan class were good tractors for furniture haulers. This one is equipped with a sleeper compartment and pulls a Trailmobile moving-van trailer for Mayflower Transit Co. of Indianapolis, Indiana. A 6V-53N Detroit Diesel is probably the power source. *Harry Patterson*

The year 1966 brought more changes, including a restyled grille and a newly designed cab and hood. A produce hauler out of Plant City, Florida, used this 70 series up and down the east coast. *Harry Patterson*

The 70 and 80 series featured both tractor and straight truck versions, available in both gas and diesel. This 1966 80 series was owned by Grant County Oil Co. Inc. of Williamstown, Kentucky. It pulled an 8,500-gallon four-compartment (3,000, 1,000, 2,000, and 2,500 gallons) Progress trailer carrying Sunoco gas products. *Progress Industries*

This is another Titan 90 with a day cab and tandem axle. It is most likely powered by an 8V-71, 318-horsepower Detroit Diesel engine. The Coca-Cola Company used this in their western operations where longer length laws applied. The Dromedary flat body was made by Master Body Works. It was used to carry extra cargo. Notice that this cab has the standard west coast mirrors similar to those of other cab-over engine tractors. *Ron Adams Collection*

For the first time ever, Chevrolet ventured into the high tilt-cab, cab-over engine model with the Titan 90. It was available in day and sleeper cab, tandem- and single-axle drive, and gas and diesel. This Titan 90 is a single axle with a day cab pulling a Fruehauf trailer. Notice the differently-styled mirrors compared to those of other cab-over engine tractors. *GMC (Chevrolet Division)*

Diamond T, like all other truck manufacturers, produced cab-over engine models. They used two styles of cabs on the 931 C models. This 931 C has a day cab with what is known as a bustle back sleeper cab. This one ran over the road for Navajo Freight Lines Inc. of Denver, Colorado, pulling a Timpte van trailer. *Diamond T*

CHAPTER 4

DIAMOND T to DIAMOND REO

Diamond T made their first truck in 1911, and for the next 47 years they kept up with the competition. In 1958 they were taken over by the White Motor Co., which bought out REO Motor Truck Co. in 1957. Some of the models that went into the 1960s started back in the early 1950s. The 900 series was among the more popular models. The 921 C and the 931 C were cab-over models offered in a few cab variations and designed for use over the road. The 921 conventional models were the 921 B for axle back, the 921 F for axle forward, and the 921 DF for axle forward with an aluminum radiator shell. A 600 series low cab-over engine and a short-nose conventional 990 were also in production. During the mid 1960s, Diamond T saw a whole new design—the 1040 conventional design and the 1000 series cab-over engine. In 1967 the name Diamond T and REO were joined as one name and became Diamond REO.

Although the trucks wore the new name, the design stayed the same. The conventional was model 11464 and the cab-over engine was the "C" series. The new name was not only used in over the road trucking, but other industries as well, wherever trucks were needed.

Here we see another bustle back sleeper cab in a single axle version, pulling a set of Trailmobile furniture vans. Kroehler Manufacturing Co. ran this set of doubles out of their Fremont, California, plant. A total of nine different Cummins diesel engines were available in this model, from 190 to 335 horsepower. *Diamond T*

Livestock Express of South St. Paul, Minnesota, depended on this 931 C to get loads of livestock safely from farm to market in a Wilson livestock trailer. This full skirt sleeper cab was more popular than the bustle back sleeper. The gentleman's name is unknown but he poses with the rig at the Wilson Trailer Company plant in Sioux City, Iowa. *Wilson Trailer Co.*

The Diamond T was a popular truck in the southern states. H. W. Phillips and Sons Inc. of Memphis, Tennessee, used this 931 C and Lufkin Reefer trailer for hauling produce cross country. Wheelbases were available from 110 to 204 inches. This model was known as the All-American. *Ron Adams Collection*

Central Transfer and Storage Co. of Harrisburg, Pennsylvania, decided that the Diamond T 634 series tractor would fit into their operation pulling the Fruehauf trailer. The standard engine in the 634 was the DT6-170 horsepower. The optional engine was the DT6-185 horsepower. The standard wheelbase was from 128 1/2 to 140 1/2 inches, with options up to 218 1/2. A sleeper cab was also available. *Diamond T*

In front of Dodger Stadium in Los Angeles is a Diamond T model 633 for the Union Oil Company. The blue and orange makes for a rather striking color scheme. *Diamond T*

Tractor Number 7 is a Diamond T model 990 pulling a Brown trailer. The owner, Midwest Solvents Co. of Atchison, Kansas, provided an add-on sleeper box for the driver's comfort. *Brian Williams*

Another model 990 owned by the Dohrn Transfer Company of Rock Island, Illinois. This model was 90 inches from the bumper to the back of the cab. It could pull a 40-foot-square nose trailer and be under the legal 50-foot length limit. The open-top trailer is a Trailmobile. Almost all the components in this model are of lightweight aluminum. *Neil Sherff*

This great looking model 921 F Diamond T was owned by T.W. Burleson and Sons Inc. of Waxihaxie, Texas. The 921 series was used for all fields of trucking. This one hauls Burleson's Honey in the Brown trailer. The available horsepower was anywhere from 185 up to 250. *Diamond T*

Another 921 F, this time in a single-axle version pulling a set of Fruehauf flatbed double trailers. Lightweight components saved an average of 100 pounds on the 4x2 and 200 pounds on the 6x4 type. Bagged cement was the load on this trip. *Diamond T*

The 931 series had basically the same appearance as the 921. The difference was that the 931 had a much bigger and higher radiator to handle much larger engines. Horsepower ranged from 195 to 335. This 931 F was owned by Dunkley Distributing Co. of Salt Lake City, Utah. It has an add-on sleeper box and is powered by a Cummins diesel. This wheelbase was typical in the western states. The reefer trailer is a Trailmobile. *Brian Williams*

The 1040 series was the first tilt hood for Diamond T. This 1040 B has a 90-degree tilting hood for easy engine access. Steel Transportation Co. Inc. of East Chicago, Indiana, spec'd this one to meet the demands of pulling iron and steel products on flatbed trailers. Steel and lightweight aluminum components were available. *Diamond T*

This 1046 F model was Diamond T's offer to the western truck operator. Nevada Rock and Sand used this 1046 F to haul a huge paydozer on the lowbed trailer. Engine outputs ranged anywhere from 220 to 380 horsepower. *Ron Adams Collection*

The 1046 F seemed to be a favorite with owner-operators. This one sported an owner-operator paint design along with an add-on sleeper. It was leased to one of the many farmers' co-ops that existed at the time. *Joe Wanchura*

The newly designed cab-over engine was the 1000-C series. Its appearance is similar to the 931 C model. Cummins, Caterpillar, and Detroit Diesels were available as the power source, both sixes and V-8s ranging from 218 to 335 horsepower. It was also offered as a straight truck for truck-trailer operations in the western states. Owner-operator Ed Hamilton had his rig leased to George Transfer and Rigging Co. Inc. of Parkton, Maryland. On this trip he pulled a Brown van trailer. *Neil Sherff*

The model 1046 F remained the same except for a restyled grille. It went from the chrome bars to a flat plate. This is when it became known as the Westerner, around 1966. The same engines were offered after the restyling. This one was decked out with an add-on sleeper box and many chrome extras. This picture was taken in 1968 at the Roadway Truck Stop in Carlisle, Pennsylvania. The tractor and Fruehauf Reefer trailer were leased to Transamerican Freight Lines Inc. of Detroit, Michigan. *Bob Parrish*

Another 1046 F that is decked out the same as the previous one. This one is owned by National Truck Service Inc. of Ider, Alabama. The trailer is a Hobbs Vanguard reefer. The picture was taken in Nogales, Arizona. *Brian Williams*

When Diamond T and REO combined, the new Diamond REO kept some of its predecessors' models' appearance. The only change was the bar on the side of the hood with the new name. This C-11464 F is owned by Riddle Truck Line of Springfield, Missouri, and pulls a Butler pneumatic bulk tank trailer. It uses the same engines and components as the previous Diamond T. *Ron Adams Collection*

Here is another model C-11464 F pulling a Wilson bottom hopper grain trailer. The wheelbases stretched up to 242 inches on the axle-forward model and 224 inches on the axle-back models. *Wilson Trailer Co.*

Another version in the "C" series is the C-11464 DB. It was also available in the axle-back setup. This fancied-up single-axle tractor shows the pride that the owner took in keeping his truck looking good. The chrome extras all stand out as does the early 1950s Dorsey moving-van trailer. This axle-back version has the Royalex tilt hood. This is a very handsome piece of equipment. *Ron Adams Collection*

Still another model C-11464 F. This time we see it with a long wheelbase, add-on sleeper, and the comforts of an air conditioner. Terminal Distributing Co. Inc. of San Antonio, Texas, was a produce hauler that ran through the southwest pulling a Utility reefer trailer. *Ron Adams Collection*

The late 1960s model C-11464 DB Diamond REO is another version with the Integral sleeper cab. This is a rather unusual cab, because everything that was made with a conventional cab used the add-on box sleeper. Diamond REO was the last to offer the integral sleeper cab. This steel hauling rig was leased to Transamerican Freight Lines Inc. of Detroit, Michigan, in their steel division. *Bob Parrish*

Diamond REO also offered cab-over engine models. This is a CO-7864, designating a sleeper cab. Horsepower ranged anywhere from 195 to 335. The wheelbase was offered as high as 206 inches. Steelcase Inc. in Grand Rapids, Michigan, fancied-up this one. It pulled a tri-axle box trailer for hauling new office furniture. *Neil Sherff*

Another C-78 model pulling a Kentucky furniture van trailer. Burnham Van Service of Columbus, Georgia, had this one leased to their fleet. *Ron Adams Collection*

Getting the "Moo Juice" from farm to dairy was the job for this C-50 series tractor and Progress BMT-600-DD 10,600-gallon train. The difference between the two cab-over engine series is that the C-78 is the sleeper cab and the C-50 is the day cab. The first trailer has a 5,100 gallon capacity and the second trailer has a 5,500 gallon capacity. The tandem dolly is a Trailco. Robert La Lone of Alma, Michigan, poses with his rig in May 1968. *Progress Industries*

Darling Freight Inc. of Grand Rapids, Michigan, is the owner of this C-50 series tractor. On this run it was pulling two 27-foot Brown trailers. This company operated in Michigan and the surrounding states. *Ron Adams Collection*

The "NC" model was introduced in 1960. This NC-900 had an NH-195 diesel engine. The owner also had the comfort of an add-on sleeper box. Wolverine Storage Co. was the agent for Allied Van Lines in the Detroit, Michigan, area. The NC-900 pulled the Highway moving-van trailer. *Neil Sherff*

CHAPTER 5

DODGE

When one manufacturer made a style change, the others seemed to follow. Dodge was no exception. Their new "D" series was introduced in 1959: the D-800, D-900, and T-900. In 1960, swing out fenders were a big hit because this design provided easy engine access for repairs.

The "NC" model was also introduced in 1960. The 900 was thc single-axle over-the-road model, while the NCT-1000 was the big tandem-axle model for mining, logging, construction, and regular over-the-road runs. A new venture for Dodge was their high-tilt cab-over engine, available in three cab sizes: a 49-inch day cab and 73- or 79-inch sleeper cabs. Dodge began offering the bigger horsepower engines as options. A noticeable change came in 1969 when Dodge went from double to single headlights. Two of the bigger trucks Dodge made were the 8x4 truck tractor for 80 ton gross weight, and 10x4 cement mixers. Dodge trucks were used in all fields of trucking and were available in all sizes, from small pickup trucks up to the Michigan centipedes.

Blatz Beer (Milwaukee's Best) is delivered by a 1962 D-600 Dodge delivery truck. A 202-horsepower V-8 engine with a four-speed transmission makes it all happen. The body was made by Remke Inc. *Dodge Inc.*

This 1961 C-600 gas job straight truck did city delivery work for Albuquerque Phoenix Express of Albuquerque, New Mexico. *Ron Adams*

Another Dodge C-600 model for 1963. It had a 178-horsepower V-8 engine with a four-speed transmission. It was available in six wheelbases from 121 to 181 inches. The owner is Peninsular Steel Co. *Dodge Trucks*

This early 1960s Dodge and Dorsey Reefer trailer was owned by Butterfield Produce Co. of Butterfield, Michigan. This is one of two designs that Dodge had at this time. *Neil Sherff*

This NC-900 diesel-powered model was leased to The Three Ivory Brothers Moving Co. in Detroit, Michigan, an agent for Allied Van Lines. The driver had the comfort and honor of sleeping in the "Ivory Tower" high rise above the cab. *Neil Sherff*

This 1964 NCT-1000 tandem-axle tractor was built for heavy-duty operations, mainly over the road. It was powered by an NH-220 Cummins diesel engine. The wheelbase ranged from 146 to 200 inches. It had swing out fenders opening a full 90 degrees for convenient engine accessibility. This unit was owned by the Worthington Steel Co. of Columbus, Ohio, and Louisville, Kentucky. *Dodge Trucks*

Here we see twenty-one brand new Dodge automobiles being hauled by none other than two Dodge NC-900 tractors. The haulaway was done by Commercial Carriers Inc. of Detroit, Michigan, established in 1928. The photo was taken on the Indiana-Ohio Turnpike or the New York State Thruway, where double 40-foot trailers were legal. *Ron Adams Collection*

Another NC-900, owned by DeRosa Transportation Co., pulling a 40-foot Highway trailer. *Ron Adams Collection*

In 1965, Dodge introduced their first high-tilt cab, cab-over engine. It was available in three cab sizes: a 49-inch day cab and either a 73- or 79-inch sleeper cab. Cummins, Caterpillar, and Detroit Diesels were available. This 900 series was owned by Kozak Distributors Inc. of Detroit, Michigan. *Dodge Trucks*

This 1000 series has the 79-inch sleeper cab. The wheelbase ranged anywhere from 122 to 200 inches. The 8V-71N, 318-horsepower Detroit Diesel was available as an option. The tractor was owned by Berman Leasing Inc. and leased by the Owens-Illinois Corp. of Toledo, Ohio. The trailer is Highway designed with two hoppers underneath for bottom unloading. *Welbridge & Bells Inc.*

In 1969, Dodge changed from double to single headlights. This NT-900 was leased to Aetna Freight Lines Inc. of Warren, Ohio, hauling steel products on this Dorsey spread-axle flatbed trailer. The standard engine in the 900 series was an NH-230 diesel. *Ron Adams Collection*

Many different combinations were pulled by Dodge trucks. Here we see a Michigan Centipede. The Michigan weight laws allowed for this many axles. A 1000 series day cab, cab-over engine pulls the two trailers to make deliveries. Notice the two flotation tires on the front axle. Mobil Oil Corporation is the owner. *Mobil Oil Corp.*

This grille style started in 1959 and lasted through 1960. This 1960 model has "Super Duty" below the Ford crest, meaning it comes from the F-850 to F-1100 series. Dairy Transport Service used their "Super Duty" Ford to pull this milk tanker trailer somewhere on U.S. 5. *Wallace Baker*

CHAPTER 6

FORD

From the company's earliest years, Ford had been among the top sellers for units produced. This held true even up into the 1960s. Ford always offered a big line of models with various options, from pickups to the highway construction models. In 1960, Ford restyled the "F" series. The tractor models were the F-750 to F-1100, C-750 to C-1100, and T-850 to T-950. A new "N" series was also introduced, followed by a new high cab-over engine "H" series in 1961. Many different gas and diesel engines were available. With all the various options, Ford offered over 1,000 models in 1963. The high rise "H" cab-over engine model was discontinued in 1966 and replaced with the new W-1000 series. This model was offered with a choice of 20 different diesel engines of various horsepower. The cab-over engine "C" model was also being produced at a good pace. A big hit for Ford came in 1969 with the introduction of the new "L" series—also known as the Louisville model—from the Louisville, Kentucky plant. This model offered a total of 30 different gas and diesel engines. All the different models and options put Ford very high on the number of units produced through the 1960s.

In 1961, Ford made changes in the design of the hood, fenders, and grille. This F-1000 has a wheelbase of 194 inches. The engine is a 477 SD V-8 with a Spicer 6852 G main transmission. Publix Super Markets of Lakeland, Florida, had a fleet of these F-1000s to pull Miller trailers in the fleet. *Ford-Houston Branch*

This T-950 was a demonstrator to New Enterprise Equipment and Supply in New Enterprise, Pennsylvania. The account was pleased enough to purchase three additional T-950s for hauling 25-ton concrete bridge beams. They rode on a 176-inch wheelbase powered by a Ford 534 SD V-8 engine with Spicer five-speed main and four-speed auxiliary transmissions. *Ford-Houston Branch*

This mid- to late-1960s F-800 was photographed at the Philadelphia Produce Center. Raymond Bolzan of Great Meadows, New Jersey, was a produce grower and hauler on the east coast. The F-800 had a standard 361-ci HD V-8 engine and five-speed transmission. A 391-ci HD V-8 engine with an eight-speed or transmatic transmission was an option. *Bob Parrish*

One of the rigs in the Minute Maid Company fleet was this F-1000 single-axle Ford, pulling a late 1950s Great Dane trailer. It was powered by a 253-horsepower 477 super duty V-8 engine on a wheelbase from 146 to 212 inches. The grille guard protected the grille and headlights from damage. *Interstate Photographers*

The "C" series was a carry-over from the 1950s. Although there is no series number description, this CT was set up for heavy work. A Cummins diesel was probably the power source. It is decked out with all the necessary equipment to do heavy hauling for C & H Transportation Co. Inc. of Dallas, Texas, a nationwide heavy hauler. The year 1961 marked the changeover from double to single headlights. *Harry Patterson*

Beverage Leaseway Inc. of Philadelphia, Pennsylvania, leased this CT-850 to C. Schmidt & Sons Inc., also of Philadelphia, to haul their beer products in the 24-foot Brown van body. The power source was Ford's 477-V4 SD engine with a Spicer 6352 transmission. This rig has a 201-inch wheelbase on 10.00x20 tires. The colors were red and gold, with white stripe. *Ford-Houston Branch*

Ford's high cab-over engine model, known as the "H" series, was introduced in 1961. This HTD-950 was owned by Umthun Trucking Co. of Eagle Grove, Iowa. The 158-inch wheelbase tractor rode on 10.00x22 inch tires. It was powered by a Cummins NH-220 diesel engine with a Fuller RA-96 10-speed transmission. Umthun hauled general freight and grain products in trailers such as this Dorsey grain trailer. *Ford-Houston Branch*

Although the model number of this "H" series is unknown, we do know that it is leased to Mayflower Transit Co. of Indianapolis, Indiana. It is a single-axle version with a sleeper cab pulling a Trailmobile furniture trailer. *Brian Williams*

This single-axle sleeper cab version is an H-1000-D from 1963. It is owned by the Athens Stove Works Co. in Athens, Tennessee, and hauls ranges in the Fruehauf trailer. The fiberglass sleeper compartment measured 25 inches. Diesel power ranged from 180 to 265 horsepower. *Kenneth B. Kile*

Inter-mixing was a common practice among some truck manufacturers. In the case of the "H" series, the low "C" model cab was used, with a grille that looks similar to that of the "F" series. This 950 was owned by Walnut Grove Products Co. Inc. of Atlantic, Iowa, and pulled a Wilson grain trailer. A diesel engine was the power source. *Ron Adams Collection*

The year 1963 brought another new model to the Ford line—the "N" series, which was Ford's short conventional, fitted with the "F" series cab. This N-1000-D is powered by a Cummins NHE-195 diesel and a Fuller transmission. It is owned by Dairy Transport Company of Somerville, Massachusetts, hauling milk products in the New England area. Their appreciation for a similar demonstrator tractor convinced Dairy Transport Co. to switch from gas to diesel power. *Ford-Houston Branch*

A lot of car haulers favored Fords. Associated Transports Inc. had this N-1000 diesel hauling new cars throughout the west. This was a western-type car-hauling rig. Notice the three U.S. Air Force Chevrolet station wagons that make up part of the load. *Brian Williams*

Associated Transport Inc. of New York City, New York, had a number of 950 model Fords in their fleet. (This company is not connected to the Associated Transport Inc. in the previous car hauler photo.) The standard engine was an NHE-195 Cummins diesel, with Detroit Diesels as an option. This company covered almost all the eastern half of the United States. *Ron Adams Collection*

After six years of the "H" series, the new "W" series was introduced in 1966. Power was offered from 20 different diesel engines: 12 Cummins, 6 Detroit Diesel, and 2 Caterpillars. This one is a W-1000 owned by Cleadous Naylor of Dunn, North Carolina. Produce was most likely the commodity hauled in the Trailmobile reefer trailer. The sleeper cab measured 82 inches. *Ron Adams Collection*

Here we see another W-1000, but this time with a day cab. This 1969 model, owned by Western Freight Lines Ltd. of Chatham, Ontario, Canada, pulls a Trailmobile trailer. *Western Freight Lines Ltd.*

Designed in the late 1960s, the new Louisville Line "L" series was introduced in 1970 to replace the "F" series. Here we see the Ford cheerleaders cheering for the new "L." The standard engine was the V175-V8. A V200, V225, and 6V-553N were options. Total Ford truck sales for 1970 was 682,789 units, of which 90,200 were heavy duty, including 24,981 of the new Louisville Line. *Ford Motor Co.*

The new Louisville Line also included the set back axle on the LTS. This was mainly the construction model, rated at 64,000 pounds. This LTS-9000 is shown with a concrete mixer body. Notice the 18.00x22.5 balloon tires on the front axle. The standard engine in this model was an NH-230, with a five-speed transmission. *Ford Motor Co.*

A WFT-6364 of the early 1960s parked in front of Utah's snow capped mountains. The tanker carrier W. S. Hatch Co. of Woods Cross, Utah, owned the truck-trailer combo. They used this body and pull trailer for transporting compressed gas. Hatch was a long-time user of Freightliner trucks in several different configurations to run the eleven western states. *W. S. Hatch Co.*

CHAPTER 7

FREIGHTLINER

Since the beginning of trucks, virtually all manufacturers had a line of different models in both the conventional and cab-over engine types. Freightliner was the only manufacturer that did not offer a conventional tractor in their model line until the mid-1970s. The company made a few conventionals in the mid-1940s, but only for use in the Consolidated Freightways Fleet, and they were not offered to the public. Freightliner started with the bubblenose, but in 1952 made their first flat-front cab-over engine. Their first tilt cab was produced in 1958. The appearance of the Freightliner changed very little through the 1960s. Freightliner production in 1961 was 1,242 units. There was a plant in Pomona, California, and around 1965, the company opened a new plant in Indianapolis, Indiana. Truck sales were increasing in the eastern United States and the new tilt cab further increased Freightliner sales. Customers included both companies and owner-operators. Freightliner had a first in the industry with their new 104-inch cab in 1968, known as the Van-Liner. In 1964, they introduced a half cab designed for the construction industry. The appearance of the Freightliner in these years made it one of the most popular trucks with owner-operators.

Here we have another early-1960s WFT-6364 truck-trailer combo, this one owned by P. F. Huntley of Spokane, Washington. Body and trailer are Aero-Liners, each with a Transicold reefer unit. The standard engine was an NH-220 Cummins diesel; a number of other Cummins diesel engines were available as options. Fuller and Spicer transmissions provided the gears. *Ron Adams Collection*

One of the many tractors in the fleet of Refrigerated Food Express Inc. of Boston, Massachusetts, was this WFT-7564, which pulled a Trailmobile reefer trailer. In the early 1960s, the stack was mounted in the inside right corner with the air cleaner in the inside left corner. R.F.X. covered a territory from New England and the Middle Atlantic States to the Midwest States. *Ken Hird, Refrigerated Food Express, Inc.*

The Western States had a number of different rig combinations. This type was known as a Dromedary, which used a short body between the cab and trailer for hauling extra freight. Many Freightliners in the west were set up like this WFT-7564 long wheelbase. Wholesale Floor Coverings of Seattle, Washington, hooked this 40-foot Brown trailer to the Freightliner tractor to haul their products. The Dromedary body was made by Trailmobile. *Ron Adams Collection*

Here we have another WFT-7564 on a cattle drive. Freightliners, like many other makes, were used in all segments of trucking. Five different cab sizes were available: 51, 63, 72, 75, or 81 inch. This tractor pulled a Hobbs Possom Belly livestock trailer. *Ron Adams Collection*

The western states always had all types of different combinations. This White-Freightliner is set up as a Dromedary flat. Johnny Teresi of Lodi, California, was the owner of unit number one. The piece of machinery is loaded on a Fruehauf flatbed trailer. Notice the three green cab lights. *Brian Williams*

Another rig owned by Johnny Teresi of Lodi, California, was this White-Freightliner truck-trailer combo. Truck-trailers were probably the third most popular combination in the western states, behind double trailers and van trailers. This cab also has three green lights. The extra chrome shows off this rig very nicely. *Brian Williams*

Freightliners were the choice for many western carriers, but owner-operators were also among the customers. Seen here is a western-style long-wheelbase tractor pulling a Fruehauf Reefer trailer. W. J. Digby Inc. of Denver, Colorado, employed this sharp-looking rig to haul refrigerated products in the western states. During the 1960s, the chrome options on this rig eventually became standard equipment. *Ron Adams Collection*

Freightliner also did some special cab design engineering at the customer's request, as seen here with this sky high sleeper. The operator enters the sleeper through the inside roof of the cab. Livestock Transport Co. of Los Angeles and El Monte, California, was the "Herd Mover" out on the range. Note the "V" type outside frame on the trailer, which was very unusual. *Stan Holtzman*

In 1964, Freightliner engineered the design of the half cab. This design was used for block hauling, dump trucks, concrete mixers, and also agricultural-type bodies. The wheelbases ranged anywhere from 175 to 235 inches. This model WFHT-6264 is equipped with balloon tires. The mixer body is made by Challenge. *Freightliner Corp.*

Short-wheelbase tractors were common for pulling double trailers. This model WFT-6342 has a 48-inch non-sleeper cab. Both trailers are outside-frame utility livestock trailers for Harris Feeding Co. in California. Notice the venting for the frontal air intake at the nameplate. This was started around 1964. *Brian Williams*

Beyond the regular trucks run in all states, Michigan also had "Centipedes." Here is one of those Centipedes, a 3-3-3-4 setup, being pulled by a 1964 Freightliner. The outfit was leased to Kramer Bros. Freight Lines Inc. of Detroit, Michigan. Freightliner trucks started making their way to the east coast in the late 1950s. *Ron Adams Collection*

During the 1960s, as the nationwide interstate system was nearing completion, both states and trucking companies were pushing for longer rigs. Oregon was one of the first states to experiment with 27-foot triples. Haney Truck Lines Inc. of Forest Grove, Oregon, used a WFT-6342 Freightliner tractor to pull this set of flatbed triples. The undercover load is unknown, although it could be lumber. Notice that at this time, Freightliner started to add a lower piece to the original bumper. *Photo Illustrators*

The Indiana and Ohio Turnpikes, the New York State Thruway, and the Massachusetts Turnpike allowed companies to run 40-foot doubles. The Federal Paperboard Company Inc. was one of the many private carriers to take advantage of double-40s operations. This Freightliner 6x4 tractor was their choice to pull Gindy 40-foot doubles. Consolidated Freightways also used many Freightliners in this same configuration. *Ron Adams Collection*

California was another state with its own special rigs. This design utilized as much available space as possible for lightweight freight while staying within the legal length and gross weight limits. This cab is about the shortest cab available for a Dromedary rig. Container Express Inc. of Los Angeles, California, owned this rig with lots of chrome extras. Notice that the Fruehauf trailer has doors at both the front and back end for loading and unloading. *Brian Williams*

During the 1970s, 1980s, and 1990s, many cab-over engine tractors had extra-long extended cabs. Freightliner was the first to introduce the extended cab. Their design was known as the "Vanliner" in 1968. It sported a 104-inch cab that was mainly designed for the owner-operator. It was known as model 10464T. A few features were a 54-inch mattress, full-length cloth compartment in the sleeper, storage compartments, and large baggage compartments. This maroon and gold Vanliner was leased to Davenport Packing Co. of Milan, Illinois. There was no shortage on the chrome extras. *Neil Sherff*

Freightliner introduced their new frontal air design in the late 1960s, featuring two circular vents on either side of the nameplate for better engine performance. Bud Doty Livestock Transportation Inc. of Atascadero, California, had this WFT-6342T pulling a set of Merritt Possum Belly livestock doubles. Both double and single headlights were available on all Freightliner trucks. *Brian Williams*

This brand new model 950 was en route to a dealer to be turned over to its happy owner. This model style continued to the mid-1960s. *Mickleth Photography*

CHAPTER 8

GMC

Like many other manufacturers, GMC had high sales because of its complete line of models from small pickups to big over-the-road tractors. In 1958 and 1959 many manufacturers introduced new models that took them into the decade of the 1960s. The "D" series cab-over engine became known as the "Crackerbox Jimmy." It was available in both day cab and sleeper cab. The 4000 model followed in 1961 along with the DFW-7000 over-the-road tractor, available in both gas and diesel. The "DL" series, the low-tilt cab, also followed, available in both tractor and straight truck versions and in gas and diesel. In 1965 the new 9500 series was introduced, available in long and short hood versions. This model was a big hit with many of the trucking companies. The long nose model was powered by the 8V-71 Detroit Diesel and the short nose by the 6V-71 engine. In 1969, GMC replaced the cab-over engine "Crackerbox" with a completely new cab design called the Astro 95. GMC had come a long way since 1911, and kept pace with the competition as the industry entered the 1960s.

GMC had a full line of trucks from pickups to big construction trucks. Here we see a 4000 model with a stake body. The standard engine was the 305 with a four-speed transmission. This was one of the new turn-of-the-decade models.

E-B Trucking Co. Inc. used this GMC V-6 gas job for hauling produce from farm to market in an early 1950s Trailmobile trailer. Although no model number is visible, I'm sure the owner was happy with rig No. 1. *Harry Patterson*

This 1961 model 4000 was owned by Progress Industries and used for hauling brand new Progress trailers to customers. The Shell Oil Co. used these trailers in their Ohio operation. The first trailer was a T-30 with a 4,050-gallon capacity and the second was a T-30A of the same capacity. The 4000 model lasted until the mid-1960s. *Progress Industries*

This model, popular with furniture haulers, was leased to Cartwright Inc. of Seattle, Washington. Its owner-operator decided to have an add-on box sleeper with dual stacks and dual air horns to give it a custom look. The trailer is a Fruehauf. *Harry Patterson*

Bertolino Produce in San Diego, California, put this DFW-7000 to work hauling produce in an early 1950s Fruehauf stainless steel trailer. It was powered by a Caterpillar diesel engine. The driver had the comfort of an add-on box sleeper, and dual stacks for a custom look. *Stan Holtzman*

Here we have another DFW-7000 with an add-on box sleeper. Buyers could choose between the 6V-71 or 8V-71 diesel engine. The available wheelbases were from 146 inches to 188 inches. The private carrier was Independent Manufacturing Co. in Kansas. *Brian Williams*

GMC also had a low-tilt cab model available in gas or diesel. This long-wheelbase tractor was set up as a Dromedary rig to use all available space for hauling light loads. Both the body and Utility trailer were set up for curbside loading and unloading. This GMC earned its keep by bringing home the bread. *Arizona Photographic Associates*

This DL-7000 poses real pretty. It is powered by a 6V-71 Detroit Diesel engine. Dual stacks and a chrome bumper give it a nice appearance while pulling this bulk feed trailer for Coast Grain Co. in Norwalk, California. *Jack Miller Photographers*

Although this low-tilt cab-over engine is a smaller tractor, the power insignia below the driver's door window tells us that it is powered by a V-12 engine. This DLW-7000 is owned by M&P Transport Ltd. of Alberta, Canada. The trailer is a Brantford rack-side open top. The license plate shows that this rig was photographed in 1962. *Brantford Trailers*

The DFW-7000 offered two engines, the 6V-71 or the 8V-71. This high-tilt cab-over engine was owned by Murphy Co-Op Transportation of Hampton, Iowa. Their business was hauling refrigerated products and also livestock, as we see here with this Chamberlin livestock trailer. Murphy had a fleet of these 7000 series tractors and also the "Cannonball" GMC cab-over engine of the 1950s. *Brian Williams*

This "air-ride" cab extended out over the front axle. This style had a huge windshield that gave good visibility. National Van Lines Inc. of Broadview, Illinois, owned this GMC tractor and model 66 series "VW" Highway moving-van trailer. *Highway Trailers*

Seeing the "D" model cab as a tractor was common, but to see it as a straight truck was somewhat rare. Jim Medlock Produce of Malvern, Arkansas, trucked a load of produce to the Philadelphia Produce Center in this big diesel-powered straight truck. A big 8V-71 Detroit Diesel was probably the choice for power. *Harry Patterson*

Most of the tractor trailers had either four or five axles. In some cases, more axles were required depending on the load. Here we see a three-axle lowboy trailer with a removable gooseneck for unloading. Moss Trucking Co. Inc. of Charlotte, North Carolina, chose a "D" model GMC tractor to handle this job with the help of an extra axle known as a "Joe Dog." The "D" cab was also available in non-sleeper design. *Jack Binch, Moss Trucking Co.*

In 1965, GMC introduced a new design in the conventional line known as the 9500 series. It was available in both short-nose and long-nose and in both gas and diesel. Both versions were tilt hoods. Wheelbases were available from 173 to 235 inches. Marvin Martin of Los Angeles, California, ordered his with an add-on sleeper box and dual stacks from the 8V-71 Detroit Diesel engine. It pulled a Timpte Reefer trailer for produce. The picture was taken in Nogales, Arizona. *Brian Williams*

Although the man in this photo is unknown, he seems to be happy with his rig. This single-axle non-sleeper DF-7000 tractor is pulling a set of Fruehauf double milk-tanker trailers for Driftwood Dairies of El Monte, California. The stack suggests it is powered by an 8V-71 Detroit Diesel engine. *Ron Adams Collection*

The long-nose 9500 series was also available in a single-axle version. Dual stacks come from the big 8V-71 Detroit Diesel engine. Smart & Final Iris Co. used their 9500 series to pull this set of Utility double trailers. *Brian Williams*

Here we have another 9500 long-nose model. A 6V-71 and an 8V-71 diesel engine were available, delivering up to 318 horsepower. Various transmissions were available with up to 20 speeds. The owner of this tractor added a custom-built sleeper box for long runs. The rig was leased to C&H Transportation Co. Inc. of Dallas, Texas, a nationwide heavy hauler. *Harry Patterson*

This is the short-nose version of the 9500 series. This version had a 195-horsepower 6-71 diesel engine as standard and a 238-horsepower 6-71 as an option. A 16-speed transmission was optional. KW Trucking Co. of Florida provided an add-on sleeper box for the driver as he hauled produce up and down the east coast in the Great Dane reefer trailer. *Harry Patterson*

In 1969 GMC replaced the previous "D" series cab-over engine with the new Astro 95 model. The same 9500-series Detroit Diesel engines were available in the Astro 95, as was a Cummins diesel. It had a lightweight aluminum tilt cab. This one belonged to Decatur Petroleum Haulers Inc. of Decatur, Alabama, and pulled a gasoline tank trailer. *Ron Adams Collection*

The Astro 95 was also available in a non-sleeper cab version. Although this is another tanker photograph, the Astro 95 was used in all fields of trucking. The Phillips 66 Company had this one working in the Michigan area pulling Michigan-style gasoline tanker trailers. Notice the big rearview mirrors versus the door mount west coast mirrors. These were probably optional. *Ron Adams Collection*

The RDF-400 series of the 1950s carried over into 1960. It was known as the West Coaster but was used in all 48 states. Walter Straus and Son of Ft. Worth, Texas, had their RDF pulling a Brown reefer trailer. Although integral sleeper cabs were available, this customer chose the add-on box sleeper. This model started as the "L" series in 1950, and then changed to the "R" series in 1953. *Neil Sherff*

CHAPTER 9

INTERNATIONAL HARVESTER

Cyrus McCormick probably never dreamed that his reaper would start a company that would become one of the biggest truck manufacturers in the world. In 1907 International started making auto buggies. In 1915, the company produced its first trucks. Throughout the years International manufactured a complete line of trucks from pickup trucks to big over-the-road truck tractors and off-highway trucks. International came into the 1960s with some famous models. One of them was the DCO-405 Emeryville. This model was very popular among trucking companies large and small, and they flooded the highways. There were a number of different models, but then the "Stars" took over. This line included the Fleetstar, Cargo Star, Paystar, Loadstar and Transtar, which replaced the models of the early 1960s. The "V" series that started in the mid 1950s made its way into the 1960s. A wide range of gas and diesel engines were offered throughout the whole lineup. We should mention that the CO-1750 city delivery cab-over engine tractor had the shortest wheelbase, at only 89 inches. They were owned by Ringsby Truck Lines Inc. of Denver, Colorado.

This model started in 1957 as the AC. In 1959 it was changed to the BC model and continued to 1961. It was a good looking tractor and was owned by Shedd-Bartush Foods Inc. of Detroit, Michigan. It pulled a Fruehauf reefer trailer. The standard diesel engine was the NH-180, though customers could select an NH-195 or NH-220. *Neil Sherff*

The "V" series was introduced in 1956 and continued for 10 years until 1965. Three gas engines were available, a V-401, V-461, and V-549. Todd Transport Co. Inc. chose a sleeper cab for their tractor, which required a wheelbase of 157 inches or more. A front guard was added for grille protection. The trailer is a Fruehauf. *Harry Patterson*

Southern Pacific Truck Service had a variety of equipment. Among their fleet was this VF-200 pulling a Converta-Flat flatbed trailer. The load was a very serious one, indicated by the "dangerous" sign on the front. The style of the "V" model did not change in its nine or ten year existence. Note that the turn signals are the same ones that were used on the AC and BC models. *Southern Pacific Truck Service*

The "CO" series, the low-tilt cab-over model, started in 1954. It was built by Diamond T under an agreement with International. Diamond T also used the same cab. It was built until 1972. This DCOF-200 was diesel powered. A few extras were added to give it somewhat of a custom look while pulling the Highway reefer trailer. Notice the gold Mack bulldog on the front. *Ron Adams Collection*

The "VCO" series started in 1956. Here we see a VCOF-190 as a tri-axle straight truck. The VCOF had a V-401 as the standard engine, with a 238-horsepower V-478 as an option. W. L. Johnson Jr. of Pompano Beach, Florida, had the produce body designed by TBC with three wheel wells to fit the chassis. *Harry Patterson*

The DCOF model started in 1956. The overall appearance stayed the same through its ten-year existence. A small change was made in 1959 from single to dual headlights. Of all the cab-over engine models that existed, this one, known as the Emeryville, was the most popular of them all. Many of them were powered by a 220 Cummins. Strickland Transportation Co. of Dallas, Texas, was a big International user. They also had many Trailmobile trailers in the fleet. *Strickland Transportation Co.*

The DCOF Emeryville was also popular with many owner-operators. This owner-operator has his rig looking good along with the Utility combination reefer and livestock trailer. This rig ran out of San Antonio, Texas, to the southwest and West Coast. *Brian Williams*

Another big user of the DCOF-405 Emeryville was Navajo Freight Lines Inc. of Denver, Colorado. The Cummins diesels in the Emeryville models ranged from 180 to 335 horsepower, while the Detroit Diesel 6V-71 and 8V-71 engines made 218 to 318 horsepower. The length of the 8V-71 was slightly shorter than the 6V-71. Navajo's tractor No. 2490 is Cummins-powered and proudly pulls the sharp-looking Fruehauf trailer saluting the Navajo Indian Nation. *Navajo Freight Lines Inc.*

The DCO-405 was the single-axle version. Two sleeper cab sizes were available, a 72 and an 80 inch. The one seen here is the 72 inch. In 1964 International went to the single inset headlights. Merle Riphagen of Artesia, California, decked out his mid-1960s Emeryville with dual air horns and stacks and pulled a set of Brown open-top bottom hopper trailers. The blue and yellow make a rather nice color scheme. *Brian Williams*

This 1964 DCOF-405 Emeryville was owned by Dick Neio of Fullerton, Pennsylvania. He leased "Miss Marilyn II" to the steel division of Transamerican Freight Lines Inc. of Detroit, Michigan. This tractor appears to have the 80-inch sleeper cab. The Emeryville and Transamerican names were popular up and down the road. *Neil Sherff*

In 1961, International released the D-400 series Emeryville. This was a new tilt hood that used the cab from the Emeryville cab-over engine model. There were nine optional engines available for this model, from 180 to 335 horsepower in the Cummins line and 218 to 290 horsepower in the Detroit Diesel line. Wheelbases ranged from 167 to 239 inches. The pictured truck is from 1961 to 1965 with "International" in writing. I.G.O. Transportation of Dinupa, California, elected to have a 72-inch sleeper cab and pulled a Utility reefer trailer. *Brian Williams*

This D-400 series was made until 1972. In 1966 it became known as the Transtar, which is also the year that "International" went to the individual letters. Seven different Cummins diesels and seven different Detroit Diesels were available. Atlantic-Richfield Company used this DF-400 truck-trailer combo to haul gasoline to local gas stations. *Brian Williams*

In 1962, International started their "Star" series. The first was the "Loadstar," which replaced the BC series and became the 1600. The wheelbases ranged from 127 to 254 inches. A wide variety of engines was available. This Loadstar was used for local distributing by the Falstaff Brewing Corp. in New Orleans, Louisiana. *Falstaff Brewing Corp.*

The Emeryville's last year was 1965, after which it was replaced by the CO-4000. The engines offered for the CO-4000 were the same as for the Emeryvilles, but the new cab was welded instead of riveted. Pacific Intermountain Express of Oakland, California, used many of these COF-4000 tractors to pull 40-foot trailers throughout their system and CO-4000 single-axle tractors to pull double trailers. The trailer is a Strick. *Brian Williams*

The Fleetstar 2000 was next in 1963, replacing the heavier "BC" models. This one, a Fleetstar A, was a totally new truck but had the same look as the Loadstar. Notice the air shield on top of the cab, which was the start of the new aerodynamic designs for better fuel mileage. It is owned by the Missouri-Pacific Truck Lines, the trucking division of the Missouri-Pacific Railroad. The trailer is a Fruehauf. *Missouri-Pacific Truck Lines Inc.*

Here we see a COF-4070-A with the 83-inch sleeper cab. It had a double-walled aluminum cab that was well insulated. There were 18 basic transmissions offered. This rig was owned by Ryder Truck Rentals and leased by the Owens-Illinois Corp. of Toledo, Ohio. The trailer is 40 feet long and has two under-mounted hoppers for bulk unloading. *Ron Adams Collection*

The new "M" series was introduced in 1964. Its main use was off-highway construction but the Ms were also used in other fields, such as oil exploration. This 6x4, owned by General Dynamics Corp., is equipped with a CMC mixer body. The cab appears to be the same type used on the RD series West Coaster tractors. *I.H.C.*

Next in the line of "Stars" was the Transtar CO-4070A, which replaced the CO-4000. This model had a custom cab that was available in three sizes: 50-inch non-sleeper, plus a 73- and 83-inch sleeper cab. It had lots more chrome than the previous cab. Some improvements were a better insulated cab, a bigger brake pedal, a simpler electrical system, and better rearview mirrors. The standard engine was International's own 230-horsepower DVT-573 V-8 diesel. There were 15 other engines available from Cummins and Detroit ranging from 218 to 335 horsepower. This one is the CO-4070 A with a 50-inch cab pulling a set of Brown doubles for A. J. Bayless, "Your Home Town Grocer," of Phoenix, Arizona. *Arizona Photographics*

Yet another "Star" was the "Cargostar," released at the turn of the decade in 1970. It was an improved CO Loadstar made at the Fort Wayne, Indiana, plant. It was offered in four- and six-cylinder models in both gas and diesel. Although they were offered in tandems for the first time, this one is a single axle. It was owned by Ehrlich-Newark Trucking and equipped with a Strick body. *Bob Parrish*

At the beginning of the 1960s, Kenworth introduced their new model, the S-922. This was Kenworth's contribution to those who pull 40-foot trailers within the 50-foot legal length. Although aerodynamic designs did not become widespread until the mid- to late-1970s, this slanted-tilt fiberglass hood is designed to improve airflow. This is a real handsome short hood tractor. *Kenworth Truck Co.*

CHAPTER 10

KENWORTH

Years ago, Kenworth trucks were unheard of and not seen in the east. In the mid-1950s they started to appear in the eastern United States, but they were still far and few between. Produced in Seattle, Washington, Kenworth was a western United States truck. It was not until the late 1950s and the early 1960s that Kenworth trucks became more common on eastern highways. The K-100 cab-over engine and the W-900 conventional were introduced around 1963 and 1964 and started flooding the eastern highways. There were more K-100s than W-900s because of the length laws. These models were available with many different types of transmissions and a variety of diesel engines with horsepower as high 475 from the 12V-71 Detroit Diesel. Kenworth was known as the Cadillac of the trucking industry. Although they were higher priced than the other makes, they became very popular with owner-operators. To keep up with demand Kenworth built a new factory in Kansas City, Missouri. In addition to their over-the-road tractors, Kenworth made off-highway machines for the construction industry—the KW-Dart trucks. Over and over, the company's trucks proved their slogan: "There's More Worth in Kenworth."

When we think of the west, we usually picture long wheelbase tractors. But short wheelbase models also served that part of the country. This one was ordered to pull a Westgold gasoline tank trailer. It is a Canadian-built tractor, which still has the butterfly hood. The Canadian-built Kenworth conventional tractors and trucks can be distinguished by their extended fender skirt and matching cover over the fuel tank. *Alex Gray Photography*

Kenworth has earned the reputation of being the Cadillac of the trucking industry. This tractor reflects that reputation, with lots of chrome to give it a real good-looking, eye-catching appearance, along with the Utility trailer. The colors of this tractor are red and light blue. *Kenworth Truck Co.*

Here we have another truck with the butterfly hood style. This existed until around 1962. It was also around this time that air conditioning was becoming a regular feature on a lot of trucks, as the huge unit on top of the cab demonstrates. The trailer is a Brown reefer. *Brian Williams*

Kenworths also had their part in the construction industry. Lots of concrete can be hauled in this Challenge mixer body. Single balloon tires replace the dual wheels. *Kenworth Truck Co.*

Truck-trailer setups were seen on many Kenworth trucks. This flat-body combo belonged to the Hedlund Lumber Co. in Sacramento, California. Building materials were their typical haul, but this load happens to be drums of roofing pitch. *Brian Williams*

The W-900 made its debut in 1965. The standard engine was an NH-220 Cummins. There were 22 optional engines available between Cummins, Caterpillar, and Detroit, ranging in horsepower from 180 to 380. The Gulf Oil Corp. had this narrow-nose W-900 in their fleet set up as a truck-trailer outfit. *Ron Adams Collection*

Kenworths were a big hit in the logging industry. This one is wearing all the necessary equipment for safety: a headache rack, construction fenders, and a heavy-duty grille guard. The logging trailer could be a Page & Page Inc. The diesel smoke trailing in the wind tells us that he just shifted and is on his way to deliver his load. *Englefield Studio*

This early 1960s K-525 was owned by Bigge Drayage Co. of San Leandro, California. There were four different size cabs available. This one is the 73-inch cab. The driver of this rig enjoys the comforts of an air conditioner for those warm California days. *Brian Williams*

A lot of reefer haulers used owner-operator tractors to do their hauling. Frozen Food Express Inc. of Dallas, Texas, was no exception. In the later 1950s, a lot of the owner operators started using Kenworth tractors such as this K-525 with the 86-inch cab. It looks good pulling the Fruehauf trailer. *Neil Sherff*

In 1969, Denver Chicago Trucking Co. Inc. of Denver, Colorado; Time Freight Inc. of Lubbock, Texas; and Los Angeles-Seattle Motor Express Inc. of Seattle, Washington, merged to become T.I.M.E.-DC Inc., with headquarters in Lubbock, Texas. This 1969 or later photo shows that early 1960s equipment, such as this K-521 with a 73-inch cab and Aero-Liner trailer, was still used after the merger. The year 1963 was the last year that the name Kenworth was in individual letters. *Brian Williams*

When it came to roundup time, Garibaldi Bros. Livestock Transportation of Maywood, California, was there to do the job. They used many truck-trailer combos, such as this one, to do the hauling. This K-525 was one of many in the fleet. A Merritt cattle body was mounted on the chassis and also pulled a Merritt trailer. Note that the name Kenworth is now on a bar instead of in individual letters. However, the skirt behind the front wheels on the cab was not eliminated until 1964, so this is probably a 1963 model. *Ron Adams Collection*

The K-100 series was introduced in 1963. The one big noticeable change is the elimination of the skirt behind the front wheel, which made for a lighter cab. The cab tilted a full 90 degrees. This 1965 model was a show tractor with a Fruehauf Reefer trailer. It looks like it could have been owned by Hurst Inc. of Kansas City, Missouri. *Kenworth Truck Co.*

Along with the new K-100 series came the new W-900 series. This very handsome-looking truck had a wide-front radiator tilt hood. The wide-front design was to accommodate bigger engines. The standard engine was a Cummins NH-220. Optional engines were Cummins, Detroit Diesels, and Caterpillars with horsepower up to 380. As can be seen on this 1966 model, chrome extras were plentiful. The Lufkin Reefer trailer looks very good with this very handsome tractor. *Kenworth Truck Co.*

Another W-900 of the mid-1960s, with Utility dropside reefer trailer. Zenith Transport Ltd. of Vancouver, British Columbia, hauled refrigerated products from Canada down along the Pacific Coast and the Southwest, and Mexican fruit and produce on the return trip. An air conditioner and an add-on 48-inch box sleeper provide driver comfort on this long wheelbase tractor. Notice the three green cab lights. *Brian Williams*

Kenworth also made straight trucks. Here we see a bulk feed body on a model K-100. The load of bulk Blue Ribbon Dairy Feed was delivered by Western Livestock Mineral Co. of Artesia, California. *Brian Williams*

Here we have a W-921, which is the single-axle model. It's pulling a set of bulk pneumatic tank trailers. The company is Service Transport of Bakersfield, California. *Brian Williams*

After making a delivery of Coors beer, this sharp looking W-900 and Brown reefer trailer were summoned to pick up a load of produce in either Nogales or Douglas, Arizona. The company is Border Beverages Inc. of Douglas. It seems that every distributor or trucking company that delivered or hauled Coors beer had black-and-red or red-and-white painted tractors. *Brian Williams*

Long wheelbase tractors were very popular in the west. This K-100 was owned by M & M Corporation of Ventura, California. It was in Nogales, Arizona, to receive a load of produce in its 40-foot Utility reefer trailer. Nogales was a famous town for loading Mexican produce for a return load. *Brian Williams*

The K-100 comes to us this time in this sparkling and shining truck-trailer combo. The load of steel coils is resting on a Fruehauf body and Fruehauf trailer. Notice the viewing window in the door. *Rick Manz*

The W-900 was also available in the narrow-nose version. This 1967 model was owned by Weyerhauser Inc. The Weyerhauser fleet consisted of various types of equipment. This W-923 is one of the trucks used to haul logs from forest to the mill. *Kenworth Truck Co.*

From the Land of the Big Sky Country in Montana, we have another super fantastic looking W-900 model. This 1969 model was sold by a Motor Power Equipment Co. branch. Although the name on the truck shows Big Sky Co-op of Great Falls, Montana, it is possible that the truck was owned by the Union Tractor Co. of Dillon, Montana, and leased to Big Sky. This was the color scheme of the Union Tractor Co. The products hauled in this Fruehauf reefer were raised and grown by the Big Sky Farmers and Rancher Marketing Cooperative of Montana Group. *Harry Patterson*

Another model carried over into the 1960s is the H-67. The standard engine was the End 673. Ryder Tank Lines Inc. of Greensboro, North Carolina, had a fleet of these H-67s pulling various types of tank trailers throughout the south, midwest, and the east coast. Notice that the grille on this model is the same grille that is used on the "G" series. *Mack Trucks Inc.*

CHAPTER 11

MACK

In 1901, the five Mack brothers never expected to see their venture turn into a giant of truck manufacturing. But it did just that because of their efforts and the hard work of everybody involved. There were some hard times along the way, but the mission was accomplished. From the first sightseeing bus, the Mack Senior and Mack Junior, and the famous AC Bulldog came many models over the years. The AK, the "B" series of the Thirties, the big "F" series of the 1930s and 1940s, the "EQ," the "L" series, the "H" series in the 1950s, and then the so-famous "B" series that started in 1953, and pushed its way into the 1960s. The new "F" series came in 1962 representing the newly designed cab-over engine. In the mid-1960s Mack introduced the "C" model, the "U" model, and the "R" model. Add to that the RL, the western-design conventional, and the FL, the western-design cab-over engine. The "MB" series, which was the low type cab-over engine, and the off-highway "M" series were also with the model family of the 1960s. From the 1930s on, Mack always had a big line of models to choose from. That was one of the reasons for their success from the early days right into the 1960s.

One of the many models in the "B" series that continued into the 1960s, this B-755 is powered by the End 864 V-8 Mack engine. This truck-trailer setup was owned by the Shell Oil Co. in their western operation. The driver keeps cool with the comfort of an air conditioner as he sits in the "L" series cab. *Ron Adams Collection*

The big "G" series replaced the "W" western series in 1958, but was only in production for four years. This Kenworth look-alike had a big cab with lots of room and a huge windshield with great visibility. This G-73 hailed from the west coast pulling a Pike trailer. *Neil Sherff*

The most famous carryover model of them all was the "B" series. This B-61 with a CAS 281 sleeper cab and Fruehauf Reefer trailer was owned by Reimer Express Lines Ltd. of Winnipeg, Manitoba, Canada. The CAS 281 sleeper cab was used on many of the B models. *Reimer Express Lines Ltd.*

Mack introduced the new "F" series in 1962 to replace the G series. This model came with the standard Mack engine, but optional engines were Cummins, Detroit Diesel, and Caterpillar. This "F" model sits at the Jersey Truck Center with a Fruehauf grain trailer waiting for a return load. *Harry Patterson*

The F-600 model was not only popular with the trucking companies but also many private carriers as well. This F-600 tractor and Trailmobile trailer are in the fleet of the Continental Sugar Co. Inc. of Chicago, Illinois. This cab also had lots of room and a large windshield giving better visibility. *Ron Adams Collection*

Here we see another B-75, this one with a CAS 281 sleeper cab. The standard engine was the End 673. M. W. Doxey of East Palatka, Florida, hauled fruit and produce through the Midwest, south, and the east coast. *Ron Adams Collection*

The most famous of all the "B" series models was the B-61. The standard engine was the End 673. This 1965 B-61T and the Heil tank trailer were in the fleet of Rollo Trucking Inc. of Keyport, New Jersey. The chrome radiator shell was optional. The standard transmission was a 10-speed direct, with a 10-speed two lever duplex or a 15-speed triplex as options. *Rollo Trucking Inc.*

The “U” series was introduced in 1964. The big feature of this model was the offset cab. The standard engine in the U-795 LST was the V-8 Maxidyne ENDT 865. The CA 361 cab had a huge windshield providing good visibility. The offset cab may have looked a little strange, but it served its purpose. *Mack Trucks*

The “F” series was introduced in 1962 but the “F” series western was introduced in 1964. The one pictured is an FL-700. Pacific Intermountain Express of Oakland, California, had many of these in both single and tandem axles. The single axles were used for doubles and the tandem axles were used for pulling 40-foot trailers. The standard engine was the EndL 673. The optional engines were Cummins or Detroit Diesels with up to 380 horsepower. *Pacific Intermountain Express*

The “R” series was next in line. It replaced the “B” series. A lot of companies used the “R” series in their fleets, but so did some owner-operators like Everett Palmer of Bentonville, Arkansas. He leased his tractor to C&H Transportation Co. Inc. of Dallas, Texas, hauling machinery and heavy equipment all over the 48 states. The add-on box sleeper was optional, as were several wheelbase lengths. *Harry Patterson*

Another “U” model pulling a Strick trailer in the fleet of Cleveland Freight Lines Inc. of Cleveland, Ohio. *Cleveland Freight Lines Inc.*

This RL-700 hails from our northern neighbors in Canada. Tractor Number 888 is nicknamed “North Star VIII.” The expensive load for West Coast Transmission contractors in Dillingham, British Columbia, is being transported by P&B Transport Ltd. of Richmond, British Columbia. Long-wheelbase tractors were plentiful in British Columbia. This tractor is equipped with a Jake brake. *P. C. Lee Photo*

Coast Motor Lines Inc. of California owned this RL-700 single-axle tractor pulling a set of double flatbed trailers. Lots of chrome and polished aluminum was available on the R Western models. The EndL 673 Thermodyne engine was standard, but 14 other Mack, Cummins, and Detroit Diesel engines were available with as much as 380 horsepower. Various transmissions were also available up to the 18-speed quadruples. *Ron Adams Collection*

This is the big "M" series Mack. These were used in mining and general construction. Load capacities ranged from 15 to 70 tons. The spec sheets suggest this truck could be an M-50X, which is the 50-ton job. The standard engine was a V-12 Cummins. The optional engine was a 16V-71 Detroit Diesel. Mack also made the M-651 XT-112 BDT, which was a 100-ton bottom dump tractor-trailer. *Mack Trucks*

The MB-600S was a six-wheel straight truck that could be set up for various kinds of hauling. The standard engine was the End 673. Judging from the rear tires, it looks like this chassis was used for some kind of construction job. *Mack Trucks Inc.*

Macks were used in almost every industry. As we see here, the logging industry was no exception. Truck No. 2 was owned by E. R. Claassen. Power comes from a Mack V-8 engine. All the necessary equipment is there to do the job. *Ron Adams Collection*

Decked out in its Sunday best, this long-wheelbase Mack Western has lots of chrome. A Dromedary body with a tool box and spread-axle flatbed trailer make for a sharp-looking rig. The available cab lengths were either 72 or 80 inches in the sleeper version. Wheelbases went as long as 254 inches. *Brian Williams*

This FL-700 was in the fleet of Arrowhead Puritas hauling drinking water. The short wheelbase tractor has the CA-50 no-sleeper cab, and the standard engine was the End 673. Optional engines, including a Cummins, pushed available horsepower as high as 380. The short wheelbase was designed to pull double trailers as seen in this photo. *Ron Adams Collection*

Although this cab style was designed for the western trucker, the eastern truck owner also took a liking to it. Henderson Produce of Cayce, South Carolina, did hauling up and down the east coast with the Great Dane Reefer trailer. *Harry Patterson*

Peterbilt trucks were always favorites with owner-operators. However, a number of trucking companies also had Peterbilts in their fleets. One of those companies was Ringsby Truck Lines Inc. of Denver, Colorado. Here we see a single-axle model 282 with a 63-inch cab. It came in handy for pulling this set of Brown reefer doubles and staying within the legal length. The standard engine was a Cummins NH-220 B, with optional choices from Detroit Diesel and Caterpillar. *Ringsby Truck Lines Inc.*

CHAPTER 12

PETERBILT

A lot of truck manufacturing companies existed into the 1930s. By the end of the 1930s, the truck crowd had thinned out. Some of them went out of business, some were bought out by others. Such was the case of Fageol Motor Trucks. Fageol quit making trucks in 1938, but the plant did not sit idle very long. In 1939, a man named Al Peterman purchased the plant for $200,000. Peterman started making trucks under the Peterbilt name using Fageol designs. The pre-war Peterbilts looked like Fageols. After World War II they took on a different look. Going into the 1950s the conventional design did not change much except for a few minor features. During the 1950s the cab-over engine model went from the Bubblenose style to the flat-front cab-over. Both the conventional and cab-over styles carried over into the 1960s. The same basic appearance continued until about 1968. This was the year they introduced the new 359 tilt-hood conventional. In 1969, the company restyled the model 352 cab-over engine. Like Kenworth and Freightliner, Peterbilts were not seen much in the eastern United States until the mid-1960s. The trucks proved themselves over the years, living up to their motto: if anything is better built, it's a Peterbilt.

Peterbilts were available in various versions. Truck-trailer combos were popular in the western states during the 1960s. This model 352 with a 73-inch sleeper cab is set up as a truck-trailer. Both body and trailer are Trailmobile. No identification is shown, but the rig is powered by a Detroit Diesel. *Detroit Diesel*

Matt J. Zaich Co. needed some help to move a Northwest shovel to a construction site. To do the job, Owl Truck Co. of Compton, California, used this model 351. Big jobs had to be done with big equipment and the Owl fleet had the ingredients to do it. *Ron Adams Collection*

A truck-trailer combo, this model 351 did its job for the California Liquid Gas Corp. of Sacramento, California. Judging by the short hood, this 351 could be from right at the turn of the decade. *Ron Adams Collection*

Fresh from the factory comes this colorful model 281 with a set of pneumatic tank bulk trailers. The 281 was the single-axle conventional model. Although we do not see any ID, I'm sure that the owner felt proud of this eye-catching rig. *Ron Adams Collection*

Here we have another 351 model at work for Western Livestock Mineral Co. of Artesia, California, hauling Blue Ribbon dairy feed to different livestock farms. *Brian Williams*

This model 352 is decked out in red, black, and white, which seemed to be the color scheme of all the companies and distributors that hauled Coors beer. Brisson Trucking Co. of Golden, Colorado, transports this load in a Brown reefer trailer. The driver enjoys the comfort of an air conditioner in this 73-inch cab, which sports a few extra chrome items. *Ron Adams Collection*

Construction sites always provided job security for aggregate haulers like Ken Knowles Trucking in California. This model 351 with Transfer dump rig has just loaded at the aggregate plant to deliver to a construction site. After the load on the trailer and body are dumped, the body of the trailer can be transferred into the body on the truck for the return trip. *Ron Adams Collection*

Roscoe Wagner Inc. of Twin Falls, Idaho, was a livestock hauler for many years. Although Wagner had a variety of different makes of trucks in the fleet, some of them were Peterbilts like this model 352 set up as a truck-trailer combination. The body and trailer are Utility and designed as a two-in-one to haul livestock one way and refrigerated products on the return. The small doors were opened for ventilation when hauling livestock. Peterbilt worked closely with the customer to give them what they wanted. *Brian Williams*

This model 282 single axle pulls double trailers for Dan Chavias and Sons of Madera, California. The undercover load on the Brown flatbed trailers is anybody's guess. *Brian Williams*

This set of Utility double-rack side flats is being pulled by a model 281. A variety of engines were available in the 281 models. The rig was owned by J. R. Young Commercial Transfer of Porterville, Lindsay, and Bakersfield, California. *Brian Williams*

One of the biggest trucking companies, and one of the biggest Peterbilt users, is Pacific Intermountain Express (PIE) of Oakland, California. PIE not only hauled dry freight but also had a few other divisions. One of those divisions was a tanker division. One of the rigs in the tanker operation that covered the western states was this model 282 that pulled a set of Fruehauf tankers. Not much chrome can be found on this 73-inch cab, but with the clean equipment that PIE ran, it still looks good. *Pacific Intermountain Express*

This great looking, mid-1960s model 351 is owned by Montana Brand Produce Co. of Salt Lake City, Utah. It is hauling a load of Black & Decker products in the Aero-Liner trailer. Lots of extra chrome and a 60-inch sleeper box were some of the options that were offered by Peterbilt, as well as the long wheelbase. The tractor is blue with a white stripe. *Neil Sherff*

By the mid-1960s, many owner-operators were buying Peterbilts. This red-and-black model 352 was leased to Trans-Cold Express Inc. of Dallas, Texas, and pulled a Hobbs reefer trailer. *Neil Sherff*

Another good looking owner-operator model 352. It was leased to Dealers Transit Inc. of Chicago, Illinois, to haul machinery and heavy equipment throughout the 48 states. Peterbilt provided many different color schemes and paint designs to suit its customers. The picture was taken at the Old Flemmings Truck Stop at Carlisle, Pennsylvania, in 1968. Notice how the headache rack and full fenders over the tandem are painted to match the cab. *Bob Parrish*

Another great-looking 352 that is hooked up to a Wilson bottom hopper grain trailer. Notice the nine cab lights—something rarely seen at that time. *Woodworth Commercial Photos*

This 351 was owned by a man named Bradley from Whittier, California. As can be seen, this tractor has many chrome extras and a 36-inch add-on box sleeper on a long-wheelbase frame. The Brown reefer trailer and the tractor make a nice combination. The photo was taken in Nogales, Arizona. *Brian Williams*

The needlenose model 351 was made into the late 1960s. This one was leased to Estes Packing Co. of Fort Worth, Texas, pulling a Utility reefer trailer. *Brian Williams*

In 1968, Peterbilt introduced a newly-designed model known as the 359. It had a much wider radiator shell to accommodate the bigger and more powerful diesel engines that were becoming available. It was Peterbilt's first tilt-hood conventional. The Coast Provision Company of Santa Fe Springs, California, ran this one cross country pulling a Brown reefer trailer. *Brian Williams*

Another change that came from Peterbilt in 1969 was the restyling of the model 352, as can be seen at the headlights. This long-wheelbase job was owned by Riteway Transport Inc. of Phoenix, Arizona, and pulled a Timpte Reefer for W. J. Digby Inc. of Denver, Colorado. Notice that a much bigger bumper was also available. This new model was known as the Pacemaker. *Ron Adams Collection*

The model 359 also became very popular with owner-operators. All the extra goodies for the 351 were also available on the 359. This one's duty was to haul livestock in the western states in the Fruehauf livestock trailer. *Joe Wanchura*

The wide-front model was also available in a single-axle version. Louis Ratzell of California had his tractor pulling a set of bottom hopper dump trailers. *Brian Williams*

This is what the new Pacemaker looked like from the factory. It was available in both two- and three-axle models. Several cab lengths were available, from 54 to 86 inches. *Transport Topics*

Introduced around 1954, the 9000 model carried over into the 1960s. This 9000 and Great Dane trailer were owned by Ryder Truck Line Inc. of Jacksonville, Florida. A Cummins NH-195 diesel engine with a 10-speed transmission provided the power. Ryder had a total of over 3,500 trucks, tractors, and trailers. Having one of the smallest cabs in the industry, it made weight savers out of these tractors. *Ryder Truck Lines Inc.*

CHAPTER 13

WHITE

What do sewing machines and White trucks have in common? Well, really nothing, except that it was the sewing machine made by Thomas White in 1859 that lead to the beginning of the White truck. The four sons of Thomas White—Thomas II, Rollin, Walter, and Windsor—ventured into making their first steam vehicle in 1900. This was the beginning of the White Motor Co. Over the years the White Motor Co. bought the following truck makers:

The White Mustang was basically the same as the 9000 model, except it was the gas model. The Mustang was used a lot in the construction industry, as can be seen here. A double channel frame was used for extra strength to handle the job. They were also engineered for flotation tires. Jahncke Service employed their Mustang as a mixer truck with a Challange-Cook mixer body. *John L. Herrmann F.R.P.S.*

Indiana Truck Corporation–mid 1930s
Studebaker Trucks
Pierce-Arrow Trucks–mid 1930s
Sterling Motor Truck Co.—1951
Autocar Trucks—1953
REO Motor Truck Co.—1957
Diamond T Truck Co.—1958

Over the years the White truck was very popular with many trucking companies. The most popular model was the 9000. It was designed to pull 35-foot trailers in 45-foot maximum-length areas and 40-foot trailers in 50-foot maximum-length areas. Its popularity took it into the 1960s along with the 3000, 4000, 5000, 7000, and the Compact. The 3000 and the 5000 were cab-over engine models that were phased out in the early to mid-1960s. Ten decades here come and goneand White trucks are still truckin' on America's highways.

The 4000 model was basically the same as the 9000 model except with a longer hood. The extra length was probably to handle bigger, more powerful engines. This owner-operator fancied up his tractor with dual stacks, air horns, bumper guides, an air conditioner and an add-on box sleeper. The rig was leased to J. H. Rose Truck Lines Inc. of Houston, Texas, for hauling heavy machinery and oilfield equipment throughout the country. *Joe Wanchura*

This 4000 model was owned by Dugdale Packing Co. of St. Joseph, Missouri. For those long trips hauling beef in the early 1960s Trailmobile reefer trailer, the driver had an add-on sleeper box. The photo was taken in Missouri. *Brian Williams*

The 3000 model made its first appearance in 1951. It went through the 1950s and carried over into the early 1960s. It was not the prettiest truck, but they proved to hold their own serving many companies beyond the thirteen or fourteen years they were produced. They worked in all fields of trucking. This one is pulling a dump trailer for H. W. Madison Co. of Medina, Ohio. *Neil Sherff*

The 3000 model was also available with a sleeper cab. Many reefer haulers had tractors with sleeper cabs because their runs were often longer than overnight. Lake Refrigerated Service Inc. of North Bergen, New Jersey, was one of those carriers. This one was diesel powered and pulled a Great Dane reefer trailer. In this wintertime shot you will notice how the vent doors were taped shut to prevent cold air from entering into the cab. *Neil Sherff*

The first high-tilt cab for White was the 5000 model introduced in 1959. The fiberglass was an advantage to many carriers because of the light weight. This one was in the fleet of North American Van Lines Inc. of Ft. Wayne, Indiana, pulling an American furniture van trailer. This was probably the first cab-over sleeper cab to have the stack riding right behind the door. *Ron Adams Collection*

Many models were offered in either a single-axle or a tandem-axle version. Here we see another single-axle version that was set up to pull double trailers on line haul runs for Pacific Intermountain Express Inc. of Oakland, California. At this time, single-axle tractors were used to pull double trailers such as this set of Trailmobile trailers. *Pacific Intermountain Express*

One of the main features that a driver wants in a cab is visibility. As you can see, visibility was no problem with this 5000 cab because of the huge windshield. This tandem-axle version was owned by Jack Hawthorne Truck Lines of Dewitt, Iowa, and pulled a Dorsey open-top livestock trailer. Because of the lightweight fiberglass cab, this rig might be able to haul a few more piggies to market than other rigs of the same type. *Neil Sherff*

In the early to mid-1960s, White restyled the 9000 model. The grille was enlarged and they went to using the Autocar cab, since Autocar was owned by White. A lot of the big carriers had many of these in their fleets, including East Texas Motor Freight Lines Inc. The Dallas, Texas, hauler covered an area from the Gulf Coast to the Great Lakes. The Lufkin trailer was one of many in the fleet. *East Texas Motor Freight Lines*

Another 9000, this one owned by Chemical Leaman Tank Lines Inc. of Downingtown, Pennsylvania. It was used throughout the fleet pulling various tank trailers such as this Butler bulk tanker. By this time many of the over-the-road tractors were tandem-axle. *Chemical Leaman Tank Lines Inc.*

Another new model in the White line was the Compact. It was designed for city delivery work. They served their purpose not only in America but also in Canada, where this truck hauled for Canadian Freightways Ltd. A Fruehauf body holds the cargo. *Canadian Freightways*

The Compact was also available in tractor form. In 1963 this tractor was doing city delivery work for Pacific Intermountain Express. The available power was the super Mustang gasoline engine, a Cummins, or Detroit Diesel engine. It had a low-set cab which made getting in and out easier. *Pacific Intermountain Express*

A new look in the model line was the restyled 4000. It still featured the old-style side-open hood. The long hood was to accommodate bigger engines. Cummins, Caterpillar, or Detroit Diesel engines were available. Spector Freight System Inc. of Chicago, Illinois, was always a big White user. The 4000 seen here pulls a brand new Fruehauf trailer. *Brian Williams*

The next high cab-over engine model to come was the 7000 model in the mid-1960s, which replaced the 5000 model and used basically the same components. Both tractor and the Trailmobile trailer were owned by the Sterling Faucet Co. in Morgantown, West Virginia. *Ron Adams Collection*

The 9000 model's new look was almost identical to the 4000 model. The only difference was the 9000's shorter hood. This tractor and Fruehauf trailer were owned by the OK Trucking Co. of Cincinnati, Ohio. *Ron Adams Collection*

Our neighbors to the north also used the 7000 model. This Cummins powered tractor is pulling Canadian-style double tank trailers from Fruehauf. Although this rig has a single-axle dolly, the engineering left enough room for a tandem-axle dolly. *Studio A. DuBuc Enrg. Provost Cartage Ltd.*

The 7000 model was used for any kind of hauling. A. Williams, location unknown, used his tractor to pull this unidentified livestock trailer. *Ron Adams Collection*

This owner-operator had his 4000 model leased to Peerless Transport Inc. and pulled a Butler tank trailer. The dual stacks indicate that the tractor was probably powered by a Detroit Diesel. *Harry Patterson*

This great looking 7000 model named "Flaming Mamie" was owned by William McDonald. He leased his rig to the steel division of the Gateway Transportation Co. Inc. of La Crosse, Wisconsin. It didn't always take a lot of chrome to make a tractor look nice. A quality paint job often did the trick. *Neil Sherff*

Crane Carrier Corp. (C.C.C.) in Tulsa, Oklahoma, manufactured off-highway trucks for the construction and oil industry. This is a drilling rig for the George E. Failing Drilling Co. to be used in the Philippine Islands. *Crane Carrier Corp.*

CHAPTER 14

MISCELLANEOUS

This fleet of four C.C.C. drilling rigs for the George E. Failing Drilling Co. is bound for the Philippine Islands to do drilling work. *Crane Carrier Corp.*

This C.C.C. is ready to serve in the construction industry hauling concrete to job sites. A Challange-Cook Bros. mixer body is doing the chore on the long-wheelbase chassis. *Crane Carrier Corp.*

This is another C.C.C. in the construction industry serving a different purpose. A load of building blocks needs to be delivered to the job site. Quite a load, but no problem for the C.C.C. *Crane Carrier Corp.*

Dump trucks were also on the menu at C.C.C. The big plus for the drivers of C.C.C. trucks is that good three-way vision. *Crane Carrier Corp.*

C.C.C. not only made off-highway trucks but also some highway tractors. Here we see a model DHT-704 that is owned by the C.C.C. and transporting new C.C.C. trucks to the dealer for customers in 1967. *Ron Adams*

FWD (Four Wheel Drive) has been making four-wheel drive trucks since the company started. This one is owned by the U.S. Air Force and was one of a fleet designed for hauling snow only. *FWD Trucks*

FWD also made some highway tractors. This was taken in 1971 at Flemmings Truck Stop in Carlisle, Pennsylvania. It was leased to Midwest Emery Freight System of Chicago, Illinois, to make a Monday morning delivery in New York. *Ron Adams*

Hayes trucks were made in Vancouver, British Columbia, Canada. Although very few served the eastern United States, a number of them were seen running up and down along the Pacific coast. This Hayes and Utility reefer trailer were owned by Del Barton of Vancouver, British Columbia. Its appearance is similar to a Diamond T. *Brian Williams*

Hayes also made cab-over engine trucks and tractors that looked like a Kenworth. Mid-West Truck Lines Ltd. of Winnipeg, Manitoba, owned this cab-over engine and Trailmobile Reefer trailer that visited the United States to make a delivery. *Neil Sherff*

From the rear the cab looks like a Freightliner, but it is actually a Hayes with a day cab. It was owned by Johnston Terminals Ltd. of Vancouver, British Columbia, and used for hauling oversize and overweight machinery. Notice the oversize tires of the tractor tandem. *Addington Photography*

Hendrickson made trucks of all types from city working trucks to the big rigging trucks. This Hendrickson did trash removal for Acme Scavenger Service with the Leach body. Although the cab is wearing the Hendrickson name, it is actually an International cab. They had some kind of a cab deal between them. It is powered by an RD-450 engine, 5x3 transmission with 11,000 pound front axle and 34,000 pound rear axle. *Hendrickson*

Hendrickson not only made all different types of trucks but also different steering and rear-end configurations. Here is a dump truck with balloon tires on all wheels. Notice the long overhang hood beyond the front axle. *Hendrickson*

This is number one of two Hendricksons that were made for Cooper-Jarrett Motor Freight Inc. of Chicago, Illinois. They were used for pulling double 40-foot trailers on the Ohio and Indiana turnpikes. They were powered by 12V-71 Detroit Diesels. A few years later Cooper-Jarrett sold them, and this one was purchased by Perry Stemen of South Gate, Michigan, and leased to the steel division of Brada-Miller Freight System of Kokomo, Indiana. The axle setup is 4-3-3-3. *Neil Sherff*

Number two is owned by Harold Buskirk. He leased his Centipede to Geo. Alger Co. in Detroit, Michigan. It is also powered by a 12V-71 Detroit Diesel. The axle setup on this unit is 4-4-2-3, for a total of 50 tires. I wonder what the ride was like. *Ron Adams Collection*

Another four-axle cab-over engine, this one owned by Stang Tank Lines of Menominee, Michigan. This twin steering axle tractor is pulling a gasoline tank trailer that looks like it could be a Fruehauf with three axles. *Hendrickson*

Buffalo, New York, was the home base of this monster truck. It is owned by Higgins Hauling and Rigging. Whatever the load may have been, it needed the assistance of this four-axle twin steering Hendrickson and all the dollies to get it to its destination. *Ron Adams Collection*

This is the same truck as in the previous photo, shown new before it left the factory. *Hendrickson*

When you look at this picture the first impression you get is that a youngster is standing in front of the off-highway quarry truck. Well, not so—look again. The unknown gentleman in this picture is dwarfed by the size of this series D-2771 KW-Dart that is used in quarries or open-pit mines. Dart trucks were made in Kansas City, Missouri. Around 1960 the Kenworth Truck Co. of Seattle, Washington, took over the Dart operations and named the trucks KW-Dart. *KW-Dart*

Marmon trucks were made in Garland, Texas. This one was owned by Doug Parker of Dearborn Heights, Michigan, and leased to Michigan Transportation Co. also of Dearborn, Michigan. The 1966 model is loaded down with steel riding on eleven axles. The "China Clipper" is ready to roll. *Detroit Diesel*

This Marmon was kept busy hauling Texas Swinging Beef and other frozen products. Being a Texas-made truck, it also pulled for a Texas-based company, Trans-Cold Express Inc. of Dallas. There were not many Marmons around since they were just starting to be produced at this time. The cargo was hauled in a Brown trailer. *Bob Parrish*

Marmon also made conventional tractors. This model CHDT was available with a Cummins, Caterpillar, or Detroit Diesel, in anywhere from straight 6 to the big V-12 configuration. It had a tilt hood, which made for easy access to service the engine. *Brian Williams*

To get a job done, the right type of equipment was needed. Here we see three Oshkosh six-wheel drive trucks. Each one has a different body designed to do the required job for the Nevada Power Co. Notice that the truck on the left displays the sign "Explosives." *Allen Photographers Inc.*

This Cat-powered Oshkosh tractor is kept busy hauling the Caterpillar bucket loader on the Hyster trailer from one job site to another. Notice the chrome stack coming up through the hood. *Ron Adams Collection*

American men were not the only ones to get drafted into the military. This all-wheel drive Oshkosh got drafted into the U.S. Army to pull a trailer that is used for hauling tanks on the war front. Notice the water pipe going from the storage tank behind the bumper up to the radiator. *Ron Adams Collection*

To prove itself, this big dual-steering Oshkosh was put to the test to see if it could handle the 80 tons placed on the Peerless lowbed trailer. Bigge Drayage Co. said "No problem." It was used for hauling big cumbersome loads, which was the service that Bigge offered. *Harry Watson Advertising*

Around the mid-1960s, REO came out with a new tilt-hood design. The tractor is owned by The Earle Equipment Company in Detroit, Michigan. It is powered by an Allis-Chalmers diesel, since Earle Equipment was an Allis-Chalmers dealer. *Ron Adams Collection*

The southwestern Arizona landscape was the background for this photo in September 1966. The REO truck with Rex mixer was owned by Columbia Ready-Mix Company. This was about the last year that the trucks wore the name REO. In 1967, the badge changed to Diamond REO. *Tucson Diesel, Inc. (REO Truck Div.)*

Denver-Chicago Trucking Co. of Denver, Colorado, had a variety of different trucks in the fleet. This 1963 photo shows a tandem-axle tractor pulling a Brown trailer. This was probably one of the bigger tractors in D-C's city delivery fleet. D-C also had some REO cab-over engine tractors in their over-the-road fleet. *Ron Adams Collection*

At first glance this tractor could be taken for a Diamond T. In fact it's an REO tractor pulling a spread-axle Trailmobile trailer. In the early and mid-1960s REO and Diamond T shared the same look on all their trucks. This cab-over engine model was owned by Lloyd Van Norden, who leased his rig to Daniels Motor Freight Inc. of Warren, Ohio. *Neil Sherff*

Here we see a large fleet of Canadian Sicard off-highway dump trucks. There is no indication of ownership, but they are probably used in some kind of mining operation. *Ron Adams Collection*

White Western Star trucks were first produced in 1968. This newly-manufactured truck came on strong over the years. This early single-axle version is pulling a set of double pneumatic bulk cement trailers for New Pueblo in Tucson, Arizona. *Ron Adams Collection*

White Western Star trucks were busy working in all industries. This one is set up as a log-hauling truck with a model 180 Savage log loader. They had a tilt hood the same as others in the industry. *Northwestern Photographics*

This colorful White Western Star is set up as a truck-trailer combination. It is owned by the Sun Lumber Company in San Pedro, California, and is used for hauling finished lumber. *Brian Williams*

Ward La France not only made fire apparatus trucks, but also commercial trucks for over-the-road use. This one was powered by a Detroit Diesel and worked for the U.S. Air Force. Notice the dual stacks and dual air breathers behind the cab of this long-wheelbase tractor. *Photo Illustrators*

With many trucking companies, regular maintenance was a must. These rigs owned by Strickland Transportation Company of Dallas, Texas, are going through a maintenance check in Strickland's shop in Little Rock, Arkansas. *Strickland Transportation Co.*

These three White model 4000s and two Mack F-600s are each pulling Fruehauf rack-side trailers. These trucks were just a small part of a fleet of trucks owned by the Union Pacific Railroad Company of Omaha, Nebraska. The truck fleet hauled materials used for track repair and other maintenance throughout the Union Pacific System. *Union Pacific Railroad*

This 1965 photograph shows four mid-1960s trucks. From left to right they are an H-67 Mack, an F-600 Mack, an HD 1000 Ford, and a Cracker Box GMC hooked to terminal trailers. The trailers consist of three Fruehaufs and one Trailmobile. The question is, are they being loaded or unloaded? *Ron Adams Collection*

The Arizona Welding Equipment Company of Phoenix, Arizona, used this Freightliner WFT-6364T for making deliveries. Its companion, the little White Compact and Utility trailer, also helped with the chores. *White Motor Co.*

Julian Martin Trucking Company was a Reefer hauler out of Batesville, Arkansas. The fleet consisted of (left to right) a Mack RL-700, two 352 Peterbilts, a CO-4000 International, and two "F" model Macks. Some of the territory they covered at the time was the New York City area and into the northeast United States. *Julian Marten Trucking*

This fleet is owned by the Cudahy Packing Company in Phoenix, Arizona. The fleet consisted of eight "C" model Ford straight trucks, one W-900 Kenworth, and one "C" model tandem-axle Ford tractor. Cudahy had several meat packing plants throughout the west. *Photography by Biers*

INDEX

Other MBI Publishing Company titles of interest:

Big Rigs of the 1950s
ISBN: 0-7603-0978-7

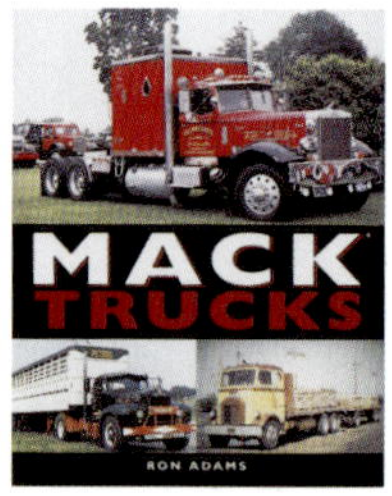

Mack Trucks
ISBN: 0-7603-1237-0

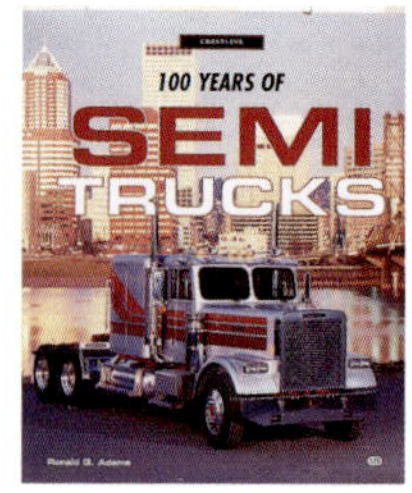

100 Years of Semi Trucks
ISBN: 0-7603-0769-5

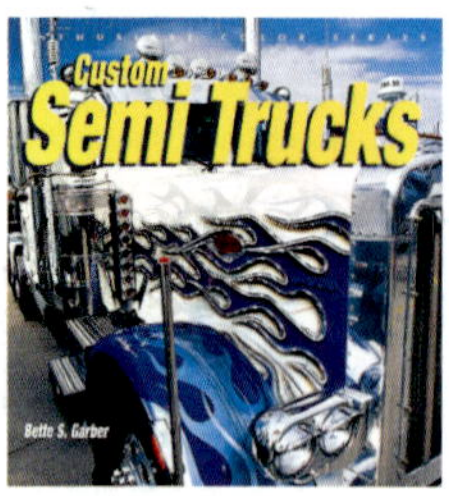

Custom Semi Trucks
ISBN: 0-7603-1459-4

Peterbilt
ISBN: 0-7603-1205-2

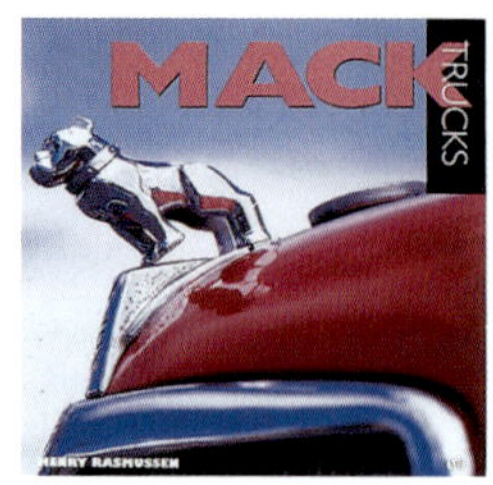

Mack Trucks
ISBN: 0-7603-1218-4

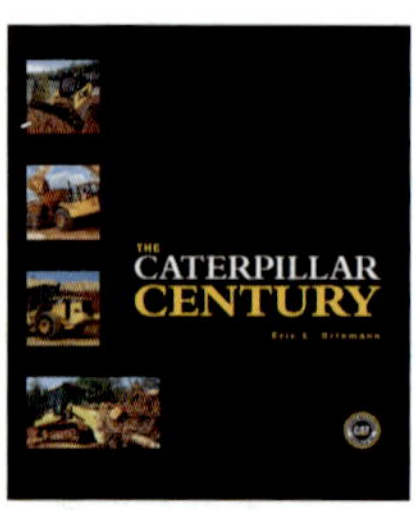

The Caterpillar Century
ISBN: 0-7603-1604-X

Dump Trucks
ISBN: 0-7603-0867-5

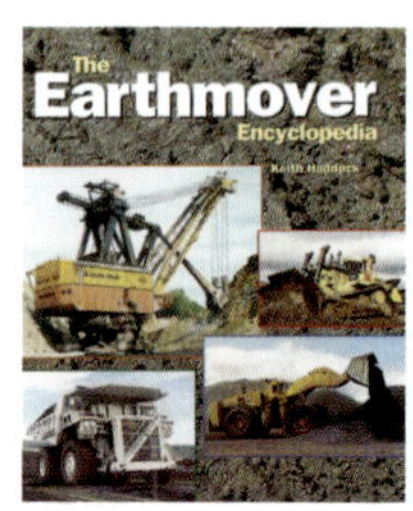

The Earthmover Encyclopedia
ISBN: 0-7603-1405-5

Find us on the internet at www.motorbooks.com 1-800-826-6600